THE GREAT OCEAN BUSINESS

BRENDA   PETER
HORSFIELD   BENNET STONE

THE GREAT
OCEAN BUSINESS

COWARD, McCANN & GEOGHEGAN INC.
NEW YORK

First American Edition 1972

Copyright © 1972 by Brenda Horsfield and Peter Bennet Stone

All rights reserved. This book, or parts thereof, may not be reproduced in any form without permission in writing from the publisher.

Library of Congress Catalog Card Number: 74-172629

PRINTED IN GREAT BRITAIN

Preface

by Sir Edward Bullard

THE GREAT DISCOVERIES OF SCIENCE HAVE AN IMPACT ON SOCIETY WHICH HAS nothing to do with their applications and for which it is difficult to account in retrospect. How, in the year 1600, could a pack of nominally Christian bishops have thought it either necessary, suitable or effective to burn Giordano Bruno for believing that the earth was not the centre of the universe and furthermore that it was moving round the sun? We can hardly conceive ourselves as thinking like them and can only haltingly explain their actions as the product of the imminent and dreaded collapse of a 1,000-year-old intellectual tradition. We may not understand them, but at least their behaviour shows a deep concern with astronomical theory.

The eighteenth-century world picture of an ideal universe and an ideal society, each running of themselves as a mechanism, a clock, almost a logical construct, strangely mirrors Newton's work on the motions of the planets. (There is a story, I forget where I found it, of a late seventeenth-century experimenter on dogs who explained that their cries did not indicate that they were being hurt but were the consequence of the laws of mechanics.) Either the whole *Weltanschauung* of the age was influenced by Newton and his friends or Newton somehow hit on the character of the succeeding century.

In our own day the successes of physics, the discovery of worlds packed within worlds like Chinese boxes and all the gadgets and inventions that have flowed from this knowledge have given us a feeling of omnipotence where any failure or distress, poverty, illness or unhappiness is thought of as an intolerable and unnecessary failure of the system of society. Clearly the developments of the great topics of natural science are of interest far beyond their immediate context; they colour an age or reflect the colours around them, it is hard to say which.

Science has been for 300 years an essential and influential part of our culture and it is important that it be presented in a way that is not only intelligible but also keeps the freshness, the drama and the clash of person-alities that accompany any scientific revolution. The subject of this book is particularly suitable for exposition in this spirit; there is no need or temptation to write down to meet a popular audience, the ideas are simple and concern everyday phenomena. But the revolution in geology, partic-ularly in the geology of the oceans, is a real revolution: in a short time we have come to look at the earth from a new point of view and many things have clicked into the new framework. Now we can teach the subject as a subject and not as a collection of facts, indeed we can refrain from teaching a lot of the facts because they follow from the new logic.

This book is written by Brenda Horsfield and Peter Stone who were not, when they started, active researchers in earth science even though Peter Stone had graduated in geology. Becoming aware of the increasingly interesting work being done in the subject in the mid-sixties, their first approach was to make popular television programmes on the results of the research 'sages'. Before long, however, they got hooked, became seriously involved and themselves started to contribute to the subject. It is one of the attractions of travelling on a rapidly accelerating bandwagon that so many people jump on and in two or three years one finds oneself among the oldest travellers and a sage in one's own right. To have been a 'populariser' and then to become seriously interested in a subject are, perhaps, the best qualifications for writing a book such as this; you can see not only the detailed scientific advances but the ramifying intellectual and practical effects and the relations to politics, industry and opinion. I hope the book finds a wide audience. It deserves to do so and can do nothing but good to the subjects it discusses.

Contents

Illustrations

Introduction

'BIG SCIENCES' ARE AREAS WHERE THE HUMAN RACE BRINGS TO BEAR ON some aspects of its environment enormous concentrations of brains and physical resources. Exactly how this happens is hard to say. Sometimes the first move is made, obviously enough, by one group of people trying to prevent another group seizing economic or military advantage. Sometimes the incentive is simply the pork-barrel or the bandwagon. Nevertheless, beneath the simple explanations there seems to lie a profound and subtle purpose, as if a whole society, invoking some fundamental faculty of co-operation, intended its own intellectual and economic evolution.

This book sets out to describe the rise of a new big science the exploitation of which will certainly cause mankind as much excitement, foreboding, danger, worry and stimulation as have space and atomic energy in their turn, and there is still no clear idea of what the ultimate cost will be. This new big science is called Oceanography. It is the whole business of getting into the sea, finding out what is there, discovering what is underneath, studying its chemistry, its physics, the dynamic systems of its movements, the nature of the different environments it offers to living things. It is also the business of moving men and equipment about 'inside' the sea, getting the capability to deal with its hazards and limitations, finding ways of getting work done at depths as great as the deep ocean floor.

What is so new about oceanography? Wasn't it already happening when Captain Cook was exploring in the 1760s, or at least since the British *Challenger* expedition set off round the world in the 1870s? Even where oceanographic exploration has overturned previous science orthodoxy, as in continental drift theory, the seminal ideas are old, certainly pre-World War II or earlier. This, however, is not the point. In space and atomic energy the same thing could have been said; the way ahead was charted by Goddard and Oberth for rocketry and by Cockcroft, Walton, Szilard and Fermi for atomic energy. But both these areas of ordinary science became big science only when whole nations backed them with significant fractions of their total national resources. Then atom bombs were actually made and men literally stood on the moon.

Perhaps the best analogy of the state of oceanography today ought to come from the first big science, atomic energy. In an atomic pile nothing very noticeable happens at first as uranium rods are added together. Although atomic reactions are going on, the new neutrons produced in the chain reactions escape. But eventually there comes a point when the pile of uranium rods is big enough and dense enough for a wholly new sort of activity to begin, the pile, as they say 'goes critical'. Temperatures go up and things begin to happen. Oceanography is now almost 'critical'.

The number of oceanographers alive and working is doubling faster than in any other sector of science; the doubling time is four years. This means that if you set up as an oceanographer tomorrow, in four years' time you will be more experienced and senior than 50 per cent of the rest of the oceanographers in the world.

Where they will lead us no one knows, yet even before the big science stage of effort and investment has been reached oceanography's most obvious goal has already been achieved. Man has already been down to the bottom of the deepest part of the ocean, 35,000 feet down in the Marianas Trench, in the dive of the bathyscaphe *Trieste* in 1960. Unlike the moon landing, this was not the summit of a multi-billion-dollar enterprise. It was a much more humble affair, well within the state of the art at the time, and well within normal science budgets.

Advance in oceanography is likely to be on a broad front with many project goals, many different teams, many different ships and submarines, from many different nations. It is most unlikely to be a great pyramid of effort as the Apollo programme has been in the United States, even under the new 'wet NASA', an oceanographic version of the National Aeronautics and Space Administration recently set up under the title of the Natural Oceanic and Atmospheric Administration, 'NOAA'. With the avowed aim of providing a 'unified approach' into the problems of the ocean and atmosphere, this agency has taken under one umbrella a whole range of different administrative bodies and departments handling everything from sport fishing to academic research into marine resources and technology.

In the advanced nations of the west people often fail to notice the beginning of major changes in the way life is lived or the way they are collectively running their affairs. Perhaps this is why it is the Japanese who are credited with discerning this new and extremely significant frontier of western life, something they were the first to call big science.

To the outside observer the obvious example of big science is space, but there are others in various states of development – particle physics, atomic energy, radio astronomy, even perhaps 'ecology'. A big science can usually be recognised by the performance goals and the national prestige it provides and it should always have an apparently noble or at least respectable purpose, capable of being expressed in some such lofty phrase as 'pushing back the boundaries of human knowledge'.

There is no short answer to the question of what shape will the world's latest big science take in the next two decades and even the long answer depends on who is talking. The world's navies for example are fundamentally concerned with one property of sea water, the way it propagates sound. Sound is the only significant energy which can travel distances of more than a few feet in the ocean and it is by sound that enemy weapons can be most easily detected. If the U.S. Navy is asked to suggest a suitable national goal the answer is the exploration of all the characteristics of ocean sound transmission.

The extractive industries give different answers. The oil industry is interested in drilling from the continental shelf and in consequential problems in ocean engineering, like how and where to store oil on the sea bed. The great metal mining companies pretend not to be interested in sea bed minerals but they are keeping a very wary eye on mid-ocean ridges and certain deposits of manganese nodules. Every time there is political trouble in countries from which non-ferrous metals are imported the mining companies offer a little more encouragement for the enthusiasts who believe in ocean mining.

The scientists are divided. A respectable academic tradition in which the British are well represented looks forward to continuing to get results by hanging bottles and corers over the ship's side much as they do now and have done for the last thirty years. A more gadget-minded younger generation, which also has its eye on commercial spin-off, foresees much more sophisticated instrumentation and a great deal of remote-controlled equipment so that ship time for the scientist himself can be used more productively. Both groups rely on heavy financial support for their research.

The major aerospace and advanced technology companies foresee ocean bed bases and large research projects to enable men to 'colonise' the deep ocean environment. They imagine a full array of underwater equivalents to bulldozers, jeeps, cranes, mechanical shovels and pipe layers. And it must be added they also see big defence installations, underwater outposts to keep an eye on the deep ocean no-man's-land.

Unlike other big sciences, oceanography is likely to involve very large numbers of people directly, especially through fishing and so-called 'aquaculture', a word increasingly used to mean any kind of food gathering from the sea or its margins. If it is the nature of oceanography to offer different futures to different men the result is no clear understanding, even for dedicated oceanographers themselves, of just where the frontier is and just what is happening there.

It is not that there is not enough information. If anything there is too much. Oceanography is expanding and changing so fast that the mere labour of collecting moderately comprehensive packages of information virtually ensures that half the words printed about it are out of date before the ink is dry.

Scientific discoveries are credited from the date of submission of papers, not from their publication and many of the learned journals are so choked up with the back-log of contributions that scientists work from circulated pre-prints and 'private communications'. So by the time a paper has appeared more work based on its evidence may already be in the pipeline and the apparently hot story is really last year's news.

The only way to keep up to date is to keep moving, catching the conferences that are honeypots to the specialists, staying on a transatlantic network that stretches between a dozen different universities and other research institutions, from the Department of Geodesy and Geophysics or the Department of Geology at Cambridge, England to the Scripps Institution of Oceanography in La Jolla, California taking in the navies of Britain, the United States and Canada on the way. Even this feverish itinerary with barely a pause to change notebooks and restock with recording tapes, would be likely to by-pass vital parts of the tangled web of discovery, initiative, finance, public relations, industrial forecasting, political sentiment and even simple human greed that are all elements in big science.

Above the national scene in any particular country looms the threat of international competition. If one nation does not develop some particular technology, some other nation will and then, like Britain and the North Sea, a country will find itself paying foreign currency to hire the means of exploiting its own resources. In the North Sea it is largely American drilling companies which have led the way in drilling for natural gas. In this case an international agreement apportioned the continental shelf between the nations extracting its resources. This is not so with fish, however, nor is it likely to be so with minerals on the ocean floor. Russian and Japanese trawlers have long fished off the Sahara, the Portuguese have fished off Newfoundland longer still, and if deep ocean manganese nodules prove to be economic they are likely to become the property of those who can get them wherever they are.

These international factors when added to all the others, to the strange and shifting image of the subject, the powerful popular appeal, the steady supply of new discoveries emerging from a daunting flow of research results beyond all hope of assimilation, make a strong brew that is very hard for governments to resist. If oceanography is not a great and growing area of profitable human activity it certainly looks like one.

Getting into the sea on the grand scale promises enormous rewards in new knowledge, new techniques and possibly sheer material profit. Equally it presents us with huge questions about what we are doing, whether this kind of effort is the best use of human resources, whether the expected returns should be proportionate to the direct investment or apportioned on some other system of values, whether the whole oceanography enterprise should be under international control not just because it is the 'last frontier' for mankind but because ungoverned technology might damage its chemistry and ecological balance beyond repair. For the ocean is not only a very large quantity of water covering a very large amount of the surface of the earth. It is also just one part of a series of complex inter-related systems that govern the total environment of the planet extending in the broadest sense to the rest of the solar system.

As oceanography gathers strength the debate that must accompany such enterprises will need all the facts that can be mustered about how things got the way they are. In this book we shall examine in some detail the rise of the particular sector of oceanography that has so far produced its most spectacular success, the exploration of the sea floor with its profound effect on the whole spectrum of the earth sciences. What has been discovered already has been so unexpected, so strange and so stirring to the imagination that it has altered the attitudes of scientists, politicians and even tax-payers to all the other possible areas of ocean research.

In 1969 an international 'Oceanology' exhibition was held in Brighton, England. Exhibitors came from all over the world with hardware, photographs, artists' impressions, bits of marine cores, depth recorder traces, displays and brochures and anything that would add flesh and contour to a speculative reality. One friend of the authors' who had long watched the scene with sceptical detachment walked through the doors in the main display room and stopped dead with astonishment – 'Good heavens!' he turned round slowly and said, 'I never thought it would happen, but I do believe they've made it. They're off! and God help the oceans.'

I

The Shaken Foundations

A CHARACTERISTIC OF BIG SCIENCE IS THAT IT SHOULD HAVE A MARKED effect on other areas of science in which it is not directly concerned. In the case of space science the unexpected consequence was the upsurging importance of solid state physics because of the engineering demand for lighter and more durable transistor switches. In the case of atomic physics the effects have turned up in the science of materials and in such odd places as welding technology or even in the new biological studies of heat pollution in lakes and rivers receiving the effluent of power stations.

Because 'oceanography' was deliberately brought into use as a general title to embrace a whole collection of different skills and studies, a great many areas of science such as physics, chemistry, meteorology, hydrodynamics, and marine biology were involved in it from the beginning. What was totally unexpected was its effect on geology.

As lately as the 1950s it was not difficult for anyone outside the discipline to dismiss classical geology as just about dead. It was clearly a subject whose days of fashionable glory had ended half a century ago. Although it had some good and dedicated workers it could never again attract the 'gunslinging' type of young talent that looks for intellectual growth areas where both discoveries and research funds will be plentiful. Yet it is the discoveries of marine geologists in their role as oceanographers that have irrigated a desert and made geology blossom into one of the most exciting fields in the whole of science today.

This dramatic revival of geology inside 10 years has been, from the point of view of public policy, almost entirely accidental. Apart from a brief rally when the world scrambled for uranium and an occasional panic call for prospectors as some indispensable industrial raw material begins to run out, geology has only rarely qualified for popular attention or government investment. In the early sixties, in the carefree years when governments and voters all believed that science still promised them a rosy future, it did look briefly as if geology was going to get a slice of the action with its own big science venture — the 'Mohole' project. This was the plan to bore a hole down to the Mohorovičić discontinuity, the base of the earth's crust where the speed of 'seismic' waves from earthquakes alters sharply and which evidently corresponds to some radical change in the composition or state of the rocks. Although this idea was put up by American geologists it had a lot of support from money-starved researchers in other countries. It was not just a piece of the mantle of the earth they hoped to see, it was the long column of marine sediments and the rocks below that would be retrieved by the drill on its way down, a record that promised to reveal the history of the primeval ocean.

The Mohole needed a lot of money. The technology did not exist to allow a ship on the surface to drill a hole in sediments and rock five miles below her keel. In the ritualised skirmishing for contracts that grows inevitably in such circumstances some real and imagined scandals were exposed that sank the whole project. Even without them it probably could not have survived the combined competition of the build-up in Vietnam and the prestige race for the moon. Now, thanks to the discoveries of oceanography, fundamental geological theory has advanced so far that the Mohole seems almost irrelevant. Some geologists even believe that the Mohorovičić discontinuity, the 'Moho', outcrops at the surface anyway, and that in certain locations, Cyprus might be one of them, you can walk past it in the sunshine. Meanwhile with a cheaper and less demanding form of deep-sea drilling the ocean sediments have been brought up after all, though by that time everybody guessed that they would at the most only contain the 'historical' record of a couple of hundred million years, probably less than one twentieth of the age of the earth.

In the years since the Mohole was dreamed up oceanographers have learned to use a whole new vocabulary that has carried the geology of the sea light-years away from the naïve past where it could be imagined that the ocean floor was a broad, flat, oozy plain bounded by gentle slopes leading upwards to the land. The whole science, they say happily, is in a state of revolution.

This particular revolution means that they now believe in an earth which is in a continuous and more or less violent state of change. However it was formed, as a separate 'cold' accretion of cosmic dust or as a fragment spinning away from the sun in a gigantic hot 'centrifuge', the earth, as it ages, is constantly altering the nature of its organisation. Living on a thin layer of crust we are all, complete with our continents and oceans, sliding across the globe on huge separate sections of rock called 'plates'. The boundaries of these plates are marked by earthquakes and volcanic mountain ranges. Each tremor and each eruption are just incidents in a system where nothing is static and all the pieces are in motion relative to each other.

Throughout remembered history the clues to all this have been there in full view of practically everybody. It may be a testimony to reckless human optimism that even in our mythologies we played down the rational sense of insecurity to go on believing that the earth was a stable platform beneath our feet. Yet its record is unnerving: earthquakes, violent floods, pillars of cloud by day, pillars of fire by night, the withdrawal of the Red Sea followed by its speedy return in the form of a devastating wave, the foundering of a great civilisation beneath the sea. It is only now, with the coming of oceanography, that we know what was going on.

To understand just how far and how fast our knowledge of the earth has come it is worth looking back to the pre-Mohole days of the early 1950s. Geologists themselves find it hard to remember what it was like in those orthodox classical times but with the help of some old editions of textbooks and dog-eared lecture notes it is possible to recapture the flavour. Some geologists even like to maintain that they knew all along that continental drift was 'right' and they always believed that there really were great convection currents in the earth's mantle to account for it (still by no means certain), but most honest recollections admit that the respectable workers frowned on these notions as heterodoxy.

This state of affairs was an accident of history, for geology in its early days was in almost a permanent revolution. Its essential foundations were laid and its first systematic work was done against a background of

religious fundamentalism – the belief that the events reported in the Old
Testament could be literally interpreted as the history of the earth. Even
when they rejected Archbishop Usher's notorious calculation – made in
1650 and based on the genealogies and ages of Methuselah and his kins-
men – that the world had been created at 9 a.m. on Sunday, October 23,
4004 B.C., people were still reluctant to let go of the idea that the flood of
Noah was, as described, a complete destruction of the first creation of the
earth with all its life forms, apart from the survivors in the ark. Fossils were
explained as the remains of the victims and if they looked sufficiently
peculiar to upset the story of a once-and-for-all creation, then they must be
'sports of nature', failed attempts at making normal modern creatures.
Short and devastating floods, floods lasting millions of years, successions of
several floods were all invoked in turn to account for the observed
geological strata and their varying collections of fossils.

Two great opposing views arose about the formation of the rocks and
led to a furious and enduring controversy. Depending on where they lived
or worked geologists were either 'Neptunists', who believed all rocks were
crystals precipitated from the sea and sediments laid down by it, or, if they
came from a volcanic part of the world, 'Vulcanists' believing that all
rocks were originally volcanic and that the sedimentary ones were their
eroded remains. Breaking away from biblical 'catastrophism' and sorting
out the relative importance of vulcanism and sedimentation gave geology
a fairly extensive experience in intellectual upheavals long before the
present excitement began.

In Britain what was to be called the 'Heroic Age' of geology began with
the enunciation by a Scottish amateur scientist, James Hutton, of the great
principle that whatever happened to the earth in the past is no different
from what is going on now. In 1795 he published his evolutionary view of
the history of the rocks in his book *The Theory of the Earth*. In 1799
William Smith, a surveyor of canals, published his method of rock
classification based on his observations that the different strata in England
could be characterised by the fossils they contained. The work of both of
them was inherited by a generation of scientists in which the professors of
geology in both Oxford and Cambridge were orthodox Church of
England clergymen, both bravely struggling to keep geology and theology
in line. In spite of the rearguard action from the religious opposition the
'evolutionary' theory of rock formation survived and flourished, particularly in the work of Charles Lyell, until it was joined by the evolu-

tionary theory of animals formulated by both Darwin and Wallace in the 1850s. In the uproar that followed the publication of *Origin of Species* in 1859 evolutionary geology quietly became respectable.

All in all the nineteenth century was a boom time in which world-wide exploration was undertaken and all the major divisions of the geological record were named, described and subdivided. In developed small countries like Britain and Switzerland every square mile of the countryside was visited by national survey bodies and mapped in some detail. Geology developed its sub-disciplines. There was palaeontology – the study of fossils, and micro-palaeontology – the study of extremely small fossils; there was mineralogy as distinct from petrology, that is to say that individual minerals were studied by different specialists from those who studied assemblies of minerals or rocks. Stratigraphy, the study of the layers of rock of different ages, spawned its own specialists like those who knew about ordinary sediments as distinct from sediments like slate which had been altered by heat or pressure.

Some geologists addressed themselves to structures and the mechanics of faulting and mountain building. Geology had its major economic successes. Oil and natural gas discoveries in particular testified to the ability of geological scientists to predict what was hidden underground. In Britain they discovered a coalfield in Kent whose existence was entirely concealed from the surface.

By World War II geology was, to all intents and purposes, a mature science. Like all mature sciences it had great controversies, ritual joustings between paladins like the battle endlessly fought between the 'liquid graniters' and the 'granitisers'. The noise of this controversy rumbled around the world but reverberated most in Edinburgh. The question concerned the origin of granite, an igneous rock which contains a relatively large amount of silica and potassium. It has the lowest melting point of all the common igneous rocks, and it occupies an important place in structural geology because the cores of mountain ranges and basement complexes are often largely comprised of it. The great question at issue was whether granite was squeezed into place as a hot liquid which then crystallised, or whether granite formed from existing sediments as a result of 'granitising fluids' which chemically altered the materials with which they came into contact. There were localities in which the granitising fluids theory was beautifully displayed with solid granite grading imperceptibly into metamorphosed sediments. In other localities granite had clearly forced its way as a liquid into cracks in the surrounding rocks, forming veins and baking the rock it had touched. Which theory you subscribed to depended on where you had been taught or where your formative field work had been done.

Departments of geology in the universities exacerbated these situations since they frequently laid territorial claim to interesting areas and it was not the done thing to trespass. Recent graduates had the restrictive choice of doing research on 'Prof's rocks' which commonly confirmed them in whatever prejudice they had unknowingly imbibed as undergraduates. Progress in a career seemed to depend on having a fortunate strike early on, finding an area of continuing interest. A classic example was that of Wager and Deer. Cruising in the thirties off Greenland, Wager happened to come on deck one morning as the ship passed Skaergaard. He noticed lines or strata of sedimentary rocks and recalled that the area was supposed to be igneous, that is comprised of rocks intruded in a molten state. He decided to take a boat ashore to investigate this disparity and found a large intrusion of igneous rock whose constituents had crystallised out. The crystals had fallen through the molten magma and had been deposited on the bottom much as sediments are deposited in water.

In a series of brilliant classical papers written over 20 years Wager and Deer described how the basic magma had cooled and fractionated into bands of rock different in composition. Could this sort of differentiation repeated on a large scale account for all the different kinds of igneous rock, from the basic black gabbros of Skye to the pale granite of the Alps? These were notable contributions to geological thinking and both men became distinguished professors at Oxford and Cambridge respectively.

The questions raised by work of this kind could not really be answered with a clear yes or no. The sort of experimental proof for theories commonly found in other sciences could not be applied in geology. It is still not possible to simulate in the laboratory the combination of temperature, pressure and time sufficient to produce geological effects. The new views won adherents and perhaps general acceptance but nothing was clinched, nothing on a large scale was solidly proved. Geologists were really like detectives looking for clues to reconstruct some past event; new theories only helped them to look in new ways or sometimes in new places. Slowly, very slowly, progress was made, but geological science had not really got the answers even to some of the questions that children ask. What causes volcanoes? Why are they where they are? How did the Alps get there and why are the Himalayas the highest mountains in the world?

Schoolboys might ask more difficult questions. Why are the Rockies and the Andes different from the Alps? What causes Ice Ages? Students asked still more difficult questions. If animals and plants cross between continents on land bridges over oceans where do land bridges go afterwards? Why has the Antarctic got coal on it? Why, if the ocean is so old, is it not more salty?

In truth the account which classical geology rendered of the major features of the earth was full of holes. The students could pick and choose between theories but only small-scale phenomena could be explained with any degree of certainty. In retrospect it is amazing that classical geology got as far as it did, considering that the rocks of two-thirds of the earth's surface are under water and a good deal of the rest is covered by glacial drift, alluvium, and dense forest.

Piecing together geological history has been a process of great complexity, carried out by generations of scientists building upon and re-evaluating the work of their predecessors. The work is slow because in general geologists can only interpret what they can get at, a convenient exposure in a river valley, a quarry that someone opens up, or a cutting made for a railway or a drain. Until recently it was only in economic geology that it was possible to go to the expense of sinking boreholes or wells to find out what is underneath the farmer's fields.

The picture of the earth's past assembled in this way was in some parts surprisingly comprehensive, but in others it was no more than plausible. It seemed to be at its least convincing in explaining the origin of such dramatic features as the great mountain chains, the Alps, the Andes, the Himalayas, or of the spectacularly thick layers of marine sedimentary rocks found on top of them, sometimes heaped into folds miles long and miles high.

The main target for European geologists was the long, high chain of the Alps. Sea shells could still be found in some places and since fossils of similar age could be seen in undisturbed strata of rock to the north, the Alpine sedimentary rocks had evidently either come from the south or been laid down where they were. The huge folds of the rocks also meant that the original area of the sediment must have been greater than the present area of the mountains. So it seemed there must have been a great sea or even a small ocean to the south. Geologists called it the Tethys, and in their struggle to explain the Alps it became as real to them as the North Sea is to any fisherman today.

So long and varied is the history of geological assault on the Alps that one waggish academic decided he could distinguish in it three great periods, as if the mound of literature were itself a geological succession. In the beginning was the Chaotic Period in which the commonest 'contributory fossil' is the indefatigable de Saussure. This man was the precursor of the typical Victorian scientist and was one of the first to climb Mont Blanc. With his large mercury barometer he ascended most of the passes of the Alps and in spite of the most diligent application he confessed to failure at the end of a lifetime's work. He could make no sense of it all, 'There is nothing constant in the Alps save their variety.'

The next was the Vertical Period in which the commonest traces of life are the work of Studer. He believed that the Alps, being vertical, got that way because they were pushed up from below. How? That was another question.

Last came the Tangential Period (although we may now be entering upon a new Sea Floor Spreading Period). The Tangential man was Albert Heim, a Swiss whose great work in the 1900s is commemorated in the name of a climbing refuge near Andermatt. With his work came the idea of 'thrusts' and 'nappes'. A great force had come from the south and pushed the Tethys sediments northwards. Those that went under the pile were metamorphosed, 'cooked' by pressure and heat; those that stayed uppermost were folded into great curves or 'anticlines' which then fell over northwards. Their tips might survive after erosion as the rootless Klippes in northern Switzerland, mountains which rest on top of debris of later age. Other folds would be sheared off completely and end up as great rafts or nappes many miles from their original roots. Sometimes the great rafts of rock had been bitten into by erosion and the floor underneath exposed. There were vast windows like the Hohe Tauern mountains where it was possible to look down through the nappes and see what the earlier stages of the Alpine 'orogeny' or mountain building had produced.

It was all a long way from the ocean but in fact, as we know now, the ocean concealed the key. At least the classical picture of the Alps did now contain the idea that sediments deposited in the sea had been heaped up and pushed over a 'foreland' (Northern Europe) by some agency usually loosely called Africa. This continent had evidently moved at least 200 miles to do the pushing. Or had it? In the thirties the battle rumbled to and fro. The Western Alpinists fought the Eastern Alpinists, the latter thought that the former's great nappes were exaggerated. Proof rested on the concept of 'facies', generally speaking on what sort of rock the nappes were made of. Not surprisingly these crushed, contorted strata had badly mangled the geologists' main source of evidence, the fossil record, so proof was virtually impossible.

In the middle thirties another theory was pushed into the picture, representing the extreme conservative view, which was highly suspicious of all this thrusting and pushing from something as solid and fixed as the continent of Africa. This was the 'gravity collapse' theory of Harrison and Falcon and it was based on observations in Iran. They proved that many of the complex structures of the Alps had been paralleled (on a less grand scale) by strata of limestone which, after being lifted up into ridge-like structures, 'anticlines', had collapsed under gravity. Some had even peeled backwards and produced thrusts and nappes. Harrison and Falcon visited the Alps and looked closely at the Falknis nappe. They interpreted its structure according to gravity collapse theory. They felt that geologists should 'examine the validity of the (rival) theory that the majority of complex structures are the direct result of lateral compression'. Stern words.

Indeed the explanation of mountain building put forward by geologists resembled the theories of economists today. These were mutually exclusive and mutually contradictory explanations for every phenomenon, and which school was uppermost depended on how highly placed were its friends and on how many books it had published. It was political but not really scientific.

What about the huge layers of the sediments? An explanation for these was more difficult than for mountain chains, one good reason being that they were less exposed. The Alps had been picked over by Europeans for nearly two centuries; they had even had tunnels driven through them for railways and roads. Other mountains in North America had also been explored and tunnelled and their various characteristics had been compared with those of the Alps, but the layers of sediment, especially during the process of formation in the sea, were accessible only by seismic methods of exploration. There were, however, a number of good examples in the geological record on land so some of their surprising features were known.

Around 1873 two American geologists, Hall and Dana, had come up with the idea of what they called 'geosynclines'. They realised that if the simpler mountain chains were analysed and the rocks put back into their original order and lined up vertically the depth would amount to 30,000 or 40,000 feet. That is far in excess of the depth of oceans today which average 13,700 feet. Furthermore in many cases all the rocks laid down in a single geosyncline appeared to have been laid down in shallow water. The only explanation was that the ocean floor must have warped downwards steadily throughout the period when the sediments were being deposited. In the original example, the Appalachians, 49,000 feet of rock accumulated over a period of 300 million years so whatever a geosyncline was due to, it was a fundamental feature of the earth's surface. Many of the details were worked out by another geologist called Haug. His picture was of long narrow troughs separating the continents whose erosion provided the beds of sediment. There were thick coarse sandy beds at the margins, finer sands further out and perhaps thin muds in the centre. Other writers elaborated on this basic idea, grafting onto it the notion of isostasy.

The principle of isostasy, first described in the eighteenth century, involves the idea that all points on the earth's surface should suffer the same gravitational pull. If any point shows a higher or lower than normal gravity it is said to show a 'gravity anomaly' and to be out of isostatic balance. To take an example, let us imagine an area covered with a 10 kilometre depth of light 'sialic' sediment (that is, sediment formed from the debris of silicon-aluminium rocks). The column of rock beneath it down to the earth's centre cannot pull as strongly as one of equal length where the top 10 kilometres is a heavier material like basalt (so called 'sima' from silicon-magnesium, the dominant elements). If the earth's crust is not rigid these imbalances should even out. Hence the idea that a great trough full of sial, a geosyncline, should be like an iceberg and should end up partially floating. So mountain chains could be explained as 'floating' geosynclines that had broken 'loose' from their anchors.

If the Alps were too grand and recent, and too complicated to let this theory take the field unchallenged, it certainly appeared to work for the ancient worn-down chains of the past made during the Hercynian mountain building period 300–400 million years ago and the even older Caledonian period.

An essential ingredient of the idea was that the sediments of the geosyncline should be compressed between two opposing forces. In the Alps the two forces were possibly Africa and Europe. But the Andes for example have several thousand miles of empty ocean on one side of them so what forms the opposing vice jaw? Some structures called island arcs are even more peculiar. Some of them, like the arc-shaped chain of the Bonin Islands in the West Pacific, are surrounded by deep ocean on both sides and are miles away from any continent. Nevertheless, island arcs have many characteristics that suggested they were mature geosynclines, chains of mountains in the making. One outstanding characteristic discovered by the Dutch scientist Felix A. Vening Meinesz is low-gravity anomalies, but folding and thrusting also help to confirm the impression of embryonic mountain chains. They frequently have volcanoes associated with them always on the concave side. Thanks to the associated volcanoes and earthquakes, they began to be called the circum-Pacific 'ring of fire'. They were evidently of considerable geological significance, stretching from the Aleutians through the Ryukyu south of Japan down towards Java and Sumatra. With all of them there was the problem of identifying the jaws of the vice that was doing the squeezing.

One cannot improve on the words of J. A. Steers who reviewed current theories on geosynclines in 1945: '. . . there is no simple explanation of these important features of the earth's surface. The time is far distant before any definite cause can be assigned to them.' In fact the time was not quite so far distant, only four years to be exact, when the first of a whole series of new discoveries and interpretations would spring up to challenge, if not immediately demolish, the whole orthodox structure of ideas – mountain building, geosynclines and all.

One area of geology in which some of the exactitude of physics was achieved was seismology. The speed of sound in a material largely depends on its density and the detailed recording of either earthquake waves or the sound of detonations began to give a better picture of underground structures. Exploration for oil had given a big boost to the science of geophysics, which now encompassed seismology and anything else in the physics locker that was relevant to geology's problems.

As geophysics brought more and more certainty, geological speculation began to be more confident. There could be little doubt for example as to the sharp differences in density between sial and sima, or between crust and mantle. The realisation came first to Andrija Mohorovičić of Zagreb when he observed the seismological records from a local earthquake in 1909. He saw that there were commonly two waves and that the greater the distance away from the earthquake the greater the separation between them. He deduced that since one wave travelled with the speed one would expect from the known density of the surface rocks, the other wave, which travelled faster, must have travelled through denser rocks underneath. The sharp difference in speeds implied a similarly sharp difference in density. According to him, this boundary occurred at a depth of 55 kilometres and it was this that became known as the Mohorovičić discontinuity or Moho. It gave substance to the old idea of 'crust' which had originally been based on the concept of a liquid earth whose outer layers had cooled and solidified as a sort of scum or slag.

Seismology also revealed that under the continents the crust could be regarded as having perhaps two layers, an upper layer of sial with a density 2·7 times that of water and a lower layer of sima with a density 3·0 or more times that of water. Under oceans there was only one layer of the dense sima. The continents were therefore recognised as being fundamentally different from the ocean floor. Furthermore it was realised that the pronounced break in the slope of the sea floor at the margins of the continental shelf corresponded to the margins of continental rock. Beyond them lay the 'abyssal plains' of the true ocean bed.

That is a very rough outline of man's understanding of the earth's major features before geology absorbed the results of ocean exploration from the 1950s onwards. It does, however, omit one huge piece of speculation which attempted to explain many of geology's unsolved problems at one stroke, though it raised even less soluble problems in doing so. This was the theory of continental drift.

Classical geology tried hard to stick close to the world as we see it. The great folds of the Alps and the gravity anomalies discovered by Vening Meinesz in the island arcs were inescapable facts. So were the huge depths of sediments and so presumably were the deep troughs continually filling up with them. The Gulf of Mexico and the North Sea today represented mid-Wales and the Appalachians 400 million years ago. When classical geology had to go to extremes such as pushing Africa into the Tethys sea to form the Alps, it was with reluctance, and men like Harrison and Falcon did their best to bring their colleagues back to the straight and narrow path. The protagonists of continental drift, on the other hand, were uninhibited dreamers. They set Africa up alongside South America, then pushed them both, together with India and Australia, down to the South Pole. Newfoundland was put next to Ireland and mountain chains were traced across great oceans and worst of all no plausible mechanism could be produced that would make any of it possible. Continents after all were enormous land masses deep and thick and heavy and millions of square miles in area. Some huge force would be needed to push them along through the earth's crust, and no such force could be found.

For anyone sympathising with these objections today it is first of all essential to get the scale right. A continent may seem a solid enough object to a traveller, just as the ocean seems very deep to the yachtsman, although in reality the amount of water on the earth's surface is in proportion to that found on a damp football, or the thickness of the blue paper on the conventional table model of the globe. The height of Mt. Everest is 10 kilometres whilst the radius of the earth is 6,375 kilometres. Looked at

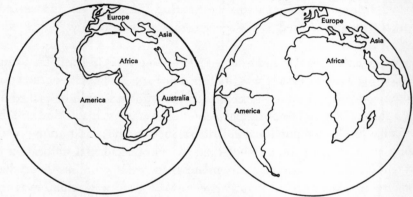

EARLIEST DIAGRAM OF CONTINENTAL DRIFT
1858 sketch by Antonio Snider-Pelligrini to explain similarities of carboniferous fossils in Europe and North America. Continents are shown before and after a 'sudden catastrophe thought to have separated them.

globally, therefore, moving a continent about is less of an upheaval than it may at first seem. Furthermore the time scales talked about from the first involved hundreds of millions of years so a moving continent is in no great hurry.

The idea didn't arrive in a hurry either. The snug fit of the outlines of South America and Africa had looked too good to be coincidence for almost as long as South America had been on the world map. By the middle of the nineteenth century when palaeontology had begun to make sense of the fossil record, similarities in the remains of European and American plants of the Carboniferous period, 300 million years ago, had touched off the argument that was to last until today. By 1885 the eminent Austrian geologist Edward Suess had entered the fray, saying that different continents with similar geological formations were fragments of an original single land mass which he called Gondwanaland, after Gondwana, a 'key' geological province in East Central India. He thought that the Indian Ocean covered part of Gondwanaland that had foundered. He also though that something similar had happened to a single Northern land mass, Atlantis – with the Atlantic covering the foundered portion. Suess was greeted with a good deal of scepticism, but in 1908 the American geologist F. D. Taylor rescued the idea of Gond-wanaland, adding the suggestion that horizontal movements, in other words continental drift, would explain the position and form of the world's fold mountains. Meanwhile, working independently, another American, Howard B. Baker, was proposing drift to account for the match of mountain ranges across the Atlantic. It is, however, the name of a meteorologist, Alfred L. Wegener, that is indissolubly linked with continental drift.

Wegener's book *Die Entstehung der Kontinents und Ozeane* (The Origin of Continents and Oceans) was published in 1915. Old copies of the book were regarded as amusing nonsense 20 years ago and could be picked up for a song. Nowadays an original second edition is a valuable find and the book has just been republished in Britain. Such are the posthumous joys of being right 50 years too soon. Wegener died of accidental suffocation in his tent on the Greenland ice cap in late 1930, well before his theory had sunk to its lowest in geological esteem. What he left behind was the proposal, far more sweeping than Suess's earlier suggestion, that *all* the continents had originally been joined together to make one enormous land mass which he called Pangaea, leaving the rest of the earth's surface covered by a single enormous ocean – Panthalassa. In a series of steps or stages the land mass had gradually broken up and the pieces had slowly 'drifted' apart into their present positions.

As a German, Wegener viewed continental drift from the ambience of northern Europe where evidence for the theory was hard to come by. His own conviction may even have been based on the feeling that explanation of the gross results of climatic change shown by geology were far easier to find by moving the continents than by envisaging extreme alterations in the behaviour of the atmosphere. Nonetheless, northern European geologists were not faced with a theoretical problem, which may explain why one of his greatest supporters was Alex L. du Toit, a South African. Paradoxically, it was du Toit who coined the name Laurasia to describe the intermittent group of the *northern* continents, but his enthusiasm for the drift theory was reinforced by his physical proximity, in his homeland,

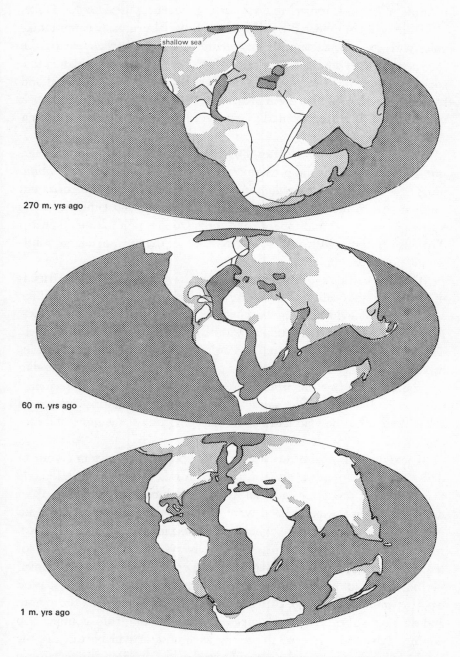

270 m. yrs ago

60 m. yrs ago

1 m. yrs ago

Wegener's three-stage reconstruction (1915) of the drifting of the continents away from a single land mass generally known as 'Pangaea'. The single great ocean covering the rest of the globe was called 'Panthalassa'. (The shaded areas on the continents denoted shallow seas.)

to some of its most remarkable supporting evidence, the Dwyka Tillite.

Tillite is a consolidated boulder clay of glacial origin. It is unmistak-able, with its assorted angular fragments of rock in a clay matrix, the fragments neither rounded like water-worn pebbles nor wind-blasted like rock fragments in desert regions. They are distinctly frost-shattered and sometimes scratched. Furthermore they often rest on smooth bed rock with characteristic scoring marks identical to those revealed by retreating Alpine glaciers today.

The Dwyka Tillites in Africa, however, date not from the last ice age but from the Permian and Carboniferous period between 234–300 million years ago. Further evidence pointing to an ice cap in south equatorial and southern Africa includes drumlins (glacial moraines) and laminated clays known as varves. Could the ice have come from a high mountain chain? The answer appeared to be no, the ice came in the form of a large conti-nental polar ice cap.

That the earth's south pole might once have been in South Africa is difficult enough to admit; much less palatable, however, is the fact that the same array of evidence can be cited for Madagascar, South America, Australia, Tasmania, Antarctica and, north of the equator, for India as well. Therefore, unless the earth's axis swung briskly about so that the polar ice cap could visit each region in turn and get in a quick ice age before moving on, the only explanation seems to be that these continents were joined together and lay far to the south of their present positions. This would allow them all to be glaciated by the spreading out of the same polar ice cap.

The Tillites have also revealed another body of evidence in favour of drift. They are found in association with fossils of two very distinctive plant genera, *Glossopteris* and *Gangamopteris*, which reached their peak of development in this period and were very abundant in the carboniferous coal measures. What is more, wherever they occur, the Tillites and the coal measures and the identical plant fossils always make the same kind of sequence. Extended at both ends to include both the preceding Devonian period, 400–350 million years ago and the following Triassic period 235–195 million years ago, they are known as the 'Gondwana succession' and du Toit and others have mapped it so assiduously from continent to continent that today it seems to provide an almost unassailable case that the southern continents have drifted and that they were originally joined into a single large unit. The conviction it carries can hardly be better expressed

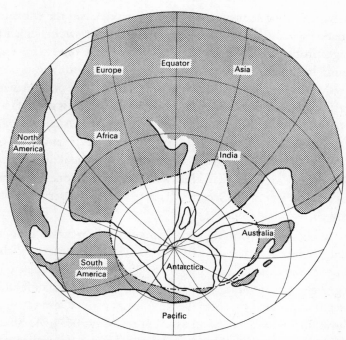

Continents assembled to explain signs of past glaciation. The white areas inside the broken line, all showing signs of ancient glaciation, especially in the Carboniferous period (which ended 280 million years ago) make one coherent polar ice cap in this reconstruction by Arthur Holmes (1944). The date and extent of the ancient glaciation is still in dispute and this model is probably an oversimplification.

than in the words of Patrick Hurley, Professor of Geology at the Massachusetts Institute of Technology:

> It is inconceivable that the complex speciation of the Gondwana plants could have evolved in the separate land masses we see today. It takes only a narrow strip of water, a few tens of miles wide at the most, to stop the spread of a diversified plant regime. The Gondwana land mass was apparently a single unit until the Mesozoic era, when it broke into separate parts. Thereafter evolution proceeded on divergent paths, leading to the biological diversity we observe today on the different continental units.

Perhaps the only reason continental drift did not originally win the day is philosophical. Geology overlooked the principle of Occam's Razor, the maxim which says 'it is vain to do with more what can be done with fewer.' If continental drift was admitted, many stratigraphical problems were solved. Only one problem was left even if it was the enormous one: what made the continents move? If continental drift were denied then the one huge problem was replaced by several lesser ones. For example a 4,000 mile long land bridge is required to account for the presence of a little freshwater reptile called Mesosaurus in both South America and South Africa during the Carboniferous period. Which is more far-fetched, a land bridge now sunk without trace into the South Atlantic; a Paleozoic Kon Tiki on which Mesosaurus might have sailed or laid eggs, or continental drift? Classical geologists, in rejecting continental drift, decided that more hypotheses of a familiar kind were better than one very unfamiliar and unpalatable one and Occam's Razor, which has served so much of science well, was laid aside.

One of the most damning early criticisms of drift was that Wegener had tried to make the two sides of the Atlantic fit together as if they were the torn edges of a card. 'To try to refit them as one refits the two parts of an irregularly torn visiting card is absurd, and it was unfortunate for Wegener that he tried to do so, because it has proved abundantly easy to show how unreasonable such an assumption is,' wrote J. A. Steers in the thirties. It is a pity that Wegener did not have a computer to do the juggling for him. With the edges of the continents defined as the 600-fathom line on the continental shelf, the computerised reassembly of the Atlantic continents published in 1964, and of the southern continents published in 1970, if they had miraculously become available at the height of the argument, would have turned him from a crank into a hero.

Here then is the picture of geology in the early 1950s. The central orthodoxy holds sway, with a 'not proven but there might be something in it' attitude to continental drift. The central orthodoxy admits the idea of continents of lightweight rock floating, but not drifting, on a substratum of heavier basaltic rock which floors the oceans as well as running beneath the continents. Beneath this two-layer crust is the mantle. The boundary between the mantle and the crust, the Moho discontinuity, is tantalisingly just out of range of drilling techniques as they then are. The old geological skills of detailed surface mapping, relative dating by means of fossils, petrological analysis of igneous rocks under the microscope, and geological prospecting with sound waves from earthquakes or explosions, are all maturely developed. Fragments, very large fragments, of the earth's surface history have been worked out, but several mysteries remain. The assumption that mountains are built from the deep troughs filled with sediment, the geosynclines, cannot be explained. Nor is it known why there are volcanoes in the middle of the ocean as well as near active mountain building areas.

Nevertheless in the fifties new techniques were showing great promise. One was radioactive dating. It had been applied as early as 1905 but its refinement and its extension to a wider variety of rocks made it much more important after World War II. With this technique it was possible to put on a suitable rock not a relative age but an absolute age expressed in years. It depends on the measurement of radioactive decay products trapped in certain minerals. In a granite for example the potassium felspar sanidine contains a certain ratio of radioactive to ordinary potassium when it crystallises from the melt. The radioactive potassium decays at a constant rate to form argon gas which remains trapped in the crystal. Measuring the amount of argon enables an age in years to be assigned to the granite intrusion.

Another new technique was the measurement of magnetic fields in the past. Here again an igneous rock can be used, since as it cools and iron minerals crystallise out they become permanently magnetised in line with the earth's magnetic field at that time. That is a useful thing to know if we assume that the earth's magnetic pole is always near the geographic pole. If the poles have been wandering with respect to the continents it should be possible to trace the effect if suitable igneous rocks can be found. A series of such rocks provide a record, a 'fossil' record, of the slow changes in the earth's magnetic field. In fact in the fifties things were turning out a little better than expected because it had been discovered that the earth's magnetic field seemed to have reversed several times and if the dates of these reversals could be established world-wide historical correlations would be made much easier. Until magnetic measurements were made it had only been possible to tell the geological time at which the rocks had been intruded or deposited. To find out the geographical latitude and longitude where the rocks had been deposited was much more problematical. One could guess that the luxuriant vegetation of the coal measures had grown in a tropical climate or that other rocks might have been formed in high polar latitudes but, since most rocks were thought to have formed under the sea, evidence which depended directly on the influence of hot and cold weather was not much use. Classical geology's interim judgment was that it was better to assume that the rocks had always been fairly near where they are found today.

POLAR WANDERING CURVES

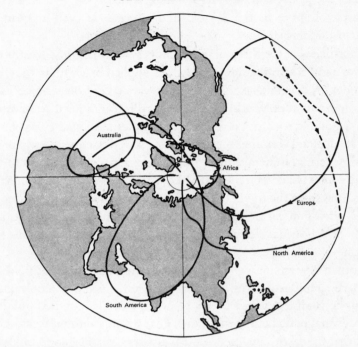

Schematic map (published 1962) of polar wandering curves compiled from palaeomagnetic studies, showing progressive 'movements' of the position of the magnetic poles, if the continents had always been in their present position.

By 1960, however, rock magnetic measurements had produced powerful new evidence for the opposition view. Willard Bascom summarised the effect:

> studies . . . lead inevitably to the conclusion that the poles and the continents have changed their relative positions. Moreover the principal part of the motion appears to have taken place since the Mesozoic (in the last 250 million years) . . . the new evidence gives strong support to the theory of continental drift.

Bascom went on to argue that what scientists needed was a stable reference point since both poles and continents move. (We know the poles also move independently of the continents.) Looking for a reference point that would meet the case he suggested the oceanic crust, since, after all, the Pacific basin is so large that one would indeed 'expect it to hold its position relatively well'.

In fact, although it is not the best simile for one of America's richest and most engaging oceanographers, Willard Bascom, like some benighted tramp seeking shelter from the rain in an old house, chose the very spot where the roof gave way first. The oceanic crust, it turned out, was anything but stable, it was moving in all directions at anything up to six centimetres a year and has been for as long as anyone can tell.

2

Starting at the Bottom

ON A CLEAR NIGHT A MAN WITH EVEN A MODERATELY POWERFUL TELESCOPE can get a good enough view of the moon to plan imaginary climbing routes down into its craters. Not even the cameras of circling astronauts give him such a good look at the earth. Quite apart from the shifting, swirling layers of cloud that always mask some part of the continents, three-quarters of the surface are permanently hidden under the dark waters of the world ocean. Yet now, along with the moon maps and the moon globes, any schoolroom can have a set of brightly coloured maps of the ocean floor and even a 3D globe showing continental shelves, abyssal plains, mountains and trenches and in fact the whole topographic detail of the ocean floor in convincing relief.

At the first glance all of these images share the same quality – they are landscapes of the mind, places that have become familiar without being visited. Look more closely and the picture of the sea floor becomes something far more extraordinary, with an extra dimension of illusion, even abstraction. For this is a landscape grander and more dramatic than anything on land yet it has never been seen by human eyes and probably never will be. It has been drawn, and in considerable detail at that, on the basis of techniques that have been justly compared to those of blind men feeling around in the water with white sticks three and a half miles long.

The need to grope for information is twofold. In the first place there is the problem of light. In water it will only travel for about 700 feet. Admittedly we can drop cameras to the very bottom of the sea and artificial sources of light to serve them but however powerful we could make them the limit of their range would remain the same. We would never, for example, *see* all of even a medium-sized marine volcano, let alone a giant like Mauna Kea in Hawaii, rising a clear 33,476 feet from the ocean floor, with its top 13,796 feet above the water surface, or a submarine mountain range as long and as high as the Alps.

In the second place there is the problem of pressure. For the men on the moon, nearly 240,000 miles from home and dangling at the end of a thin thread of communications, the environment is hostile enough – violent extremes of temperature, no humidity, no air to breathe and only a sixth of the gravity to which their bodies are adapted. All the same, for a short time at least, they can walk about and pick up rocks and plant flags with comparative freedom provided by the portable personalised environments of their spacesuits. At the bottom of the sea our most sophisticated engineering efforts can give them nothing of the sort. Although they are, at the very most, only seven miles from base their exterior environment is probably as bad as anything that men are likely to encounter anywhere else in the solar system since the obvious exception, the sun itself, hardly recommends itself as a destination. So for the foreseeable future at least the most we can hope to do down on the sea bed is move around in and peer out of vehicles that protect us from the pressures of something like 18,000 pounds per square inch.

It is with these somewhat sobering thoughts in mind that we should review the slow and stumbling efforts that have assembled all our available information about the oceans. Already as these words are being written, plans to get out to Mars, Jupiter and Venus have passed from fantasy to politics and the major problem has switched from engineering to how to persuade the taxpayers to part with the necessary money. So for most people who read newspapers and watch television it may be a surprise to learn that right here on earth the sea is not conquered, that it puts up a powerful resistance to technology and is still guarding quite a number of its oldest and darkest secrets.

Oceanographers already know this. They also know that only years of the kind of back-breaking labour that seems inescapable in any dealings with the deep ocean, together with the inspired ingenuity of a few individuals, have provided the quite extensive array of research tools now at their disposal. Today when a research ship puts to sea she may be carrying ʌighly specialised equipment and instruments to do a dozen different jobs. She can study the water itself, its temperature or chemistry or movement, she can study the creatures that live in it, she can make measurements of gravity or magnetism and on the sea floor thousands of feet below her keel sample the rocks and sediments, emplace seismographs and tide gauges and measure the flow of heat from within the earth. By measuring the time taken by sound waves travelling through them she can even explore the nature of the rocks below the sea bed itself. Yet the most dramatic discoveries that have been made so far about the oceans all stem from the methodical application of one of the oldest and simplest of recorded techniques – 'heaving the lead' – sounding the bottom with lead and line.

Before navigation became an exact art, sounding was a skill of major importance to sailors who, if they were wise, did not venture far out of sight of land. The Greek historian Herodotus has left us an account of it from the fifth century B.C. and from further back still, around the fifteenth century B.C. we have a painting from an Egyptian tomb of a sailor sounding with a long pole from the front of a riverboat on the Nile. But there is no record from the ancient world of anyone trying to make soundings in the deep sea; that had to wait for modern explorers, setting off on leisurely voyages in well-equipped ships to look for undiscovered lands and profitable trade routes. Even then, since no-one could guess that exploring the sea bottom would produce any particularly interesting results, there was no special motive behind the pioneering efforts.

The first success was scored in 1773 by an Englishman, Constantine John Phipps, later Lord Mulgrave, the captain of H.M.S. *Racehorse*. In what we now know is a basin beyond the edge of the continental shelf between Iceland and Norway he joined all his ship's lines together and lowered a 150 pound lead weight to a depth of 683 fathoms (4,098 feet). The weight sank into the bottom mud and when the line was recovered its first ten feet were found to be covered in a very fine soft blue clay. A major branch of oceanography was on its way.

Getting on for half a century later, in 1819, John Ross, making a voyage to Baffin Bay in the whaler *Isabella*, in the hope of finding the North-west Passage, used a 'deep-sea clamm' on the end of a line to make soundings and pushed the limit down to 1,000 fathoms (6,000 feet), bringing back from the bottom a sample of greenish mud. Sometimes the 'clamm' brought up animals from the depths – annelid worms and occasional crustacea and once even a starfish entangled in the lower part of the line, showing that there was life in the sea deeper than anyone believed.

Twenty years later still John Ross's nephew and second-in-command, James Clark Ross, was in command of another voyage of exploration, the Antarctic Expedition of the two ships H.M.S. *Erebus* and H.M.S. *Terror* with the mission, among other things, to determine the position of the south magnetic pole. Remembering his uncle's success James Clark Ross was determined to take deep soundings and prepared a 3,000 fathom line, fitted at regular intervals with swivels and stowed on a huge reel. In good weather the reel was fitted into a boat and the long line was run out with a 76 pound weight on the end of it. As each 100-fathom mark ran out over the gunwale the time was noted and as soon as the interval between marks lengthened it was assumed that the weight had reached the bottom. On January 3, 1840, Ross set a new sounding record with a depth of 2,425 fathoms (14,550 feet) at a position 27°26'S. by 17°29'W. in the South Atlantic.

Earlier explorers who had tried to find the bottom far out in the ocean had paid out as much as 6,000 fathoms (36,000 feet) of line, but their soundings were hopelessly exaggerated by the effects of wind and surface currents drifting their boats away from the descending line. Ross had recognised this danger and the accuracy of his sounding was achieved by dropping the line not from his ship, the *Erebus*, but from an open rowing boat held in position by another boat using its oars to correct for wind and current. Ross's whole sounding operation, including the recovery of the weight, took only four hours, but the skilled seamanship required, together with the difficulty of handling open boats in anything except calm conditions, were clearly not likely to produce such soundings in any great numbers.

Fortunately in 1845, only a quarter of a century later this time, technical advance came to the rescue of all concerned in the form of a detachable weight designed by Midshipman J. M. Brooke of the U.S. Navy. The weight was a cannon ball with a hole in the middle and when it reached the bottom, catches released it from the line which could then be hauled up more easily and faster. Brooke even designed a device to bring up a sample of the bottom. He fitted a quill or hollow tube inside the cannon ball and let the weight drive the tube into the mud, thus creating the forerunner of the modern corer.

Even with Brooke's ingenious sounder, progress in deep ocean soundings would probably have remained painfully slow, dependent on isolated opportunities to indulge pure scientific curiosity or on the interests of naval surveyors, if the invention of the 'electric telegraph' and a sudden pressure of commercial interest had not forced the pace. Then, almost by accident and concealed by the practical aim of improving communications and making money, the first really massive investment was made in the exploration of the sea bed.

In 1851 a telegraph cable was successfully laid on the bed of the English Channel between Dover and Calais and this immediately led to a scheme to lay a cable across the Atlantic. Soundings had already been made in the ocean by Lieutenant Berryman of the U.S. Navy in *Arctic* in 1856, but a lot more work had to be done before a 2,500 mile long cable would have any chance of success. So in 1857 a British naval ship, H.M.S. *Cyclops*, ran a line of soundings along the Great Circle track between Iceland and Newfoundland, the proposed route of the cable, using a hemp line. She took soundings every 50 miles of the route and each one took her six hours, with an average depth in the deep sea of two and a half miles, each sounding bringing back a bottom sample which was carried home for examination. Even in the nineteenth century ship-time was expensive and to speed up the great number of soundings required for this kind of effort the great Lord Kelvin designed a 'sounding machine' which had wires and a steam winch to speed line recovery. An early model of it was even carried by H.M.S. *Challenger* but it was not used because its drum was too weak and collapsed.

In the event setting up a transatlantic telegraph link was a bigger problem than its sponsors had believed. The first cable, laid with dramatic success in 1858 by two ships, *Agamemnon* sailing from Ireland and *Niagara* sailing from Newfoundland and meeting in mid-ocean to make the splice, failed after a few months through faulty insulation and it was several years before public confidence recovered enough from the disappointment for another attempt to be made. This time the cable broke two-thirds of the way across the ocean, and it was not until 1866 that the *Great Eastern*, Brunel's huge, controversial iron ship, paddle-wheeled and screw-propellered, did the job successfully, but only on her fifth attempt.

Meanwhile a lot had been learned about the Atlantic; it was not level as the traditional view of an ocean floor expected, but extremely uneven. At one point on the route the soundings had fallen away abruptly from 550 fathoms to 1,750 fathoms (10,500 feet) and, what in the long run turned out to be even more interesting, the deepest part was not the middle. The middle was in fact a rise which was named 'The Telegraphic Plateau'.

With the laying of this commercially important cable it became an established practice for the sea bed to be carefully surveyed in advance of the cable-laying itself. As the task of surveying was beyond the resources of the cable companies, all British cable-laying was preceded by surveys made by naval ships operated by the then Hydrographer of the Navy, Vice-Admiral Richards. This work was such an important part of his official duties that after opting for an early retirement in 1874, on the grounds that the Hydrographer should be a young and active man, Richards became Managing Director of the Telegraph Construction and Maintenance Company and over the next 20 years supervised the laying of no less than 76,000 miles of submarine cable and the development of better sounding equipment as well.

Before his retirement from the Navy, however, and before cable-laying reached its maximum intensity in the last quarter of the century, Vice-Admiral Richards had played a leading part in organising the first major effort of purely scientific exploration of the sea. This was of course the voyage round the world of H.M.S. *Challenger* between 1872 and 1875.

Challenger made soundings every 100 miles, right round the world, using old-fashioned hemp lines and a complicated piece of apparatus called an 'accumulator' involving 20 stout india-rubber ropes which acted as shock-absorbers to lessen the risk of line breakage. With this contraption she even made a record sounding of 4,500 fathoms (27,000 feet) in what we now know to be the deepest of ocean trenches, the Marianas Trench in the West Pacific. In the Atlantic she established that the central plateau was less than half as deep as the two broad troughs on either side of it. It is tempting to wonder whether, if her soundings had been closer together and the true profile of the Atlantic Basin had been discovered, complete with the mountains of the Mid-Atlantic Ridge, the whole history of oceanography would have been different.

As it was, deep-sea soundings were only one of the tasks undertaken by *Challenger*. Conventional references to the expedition as 'the birth of modern oceanography' conceal the important fact that the scientists of a century ago were spurred on by a different set of research priorities from those of today. *Challenger's* scientific staff could not even guess at the geological and geophysical revelations that preoccupy so many of their modern successors. Far clearer to them was the need for extensive observations of water and air to combat the dubious popular theories of ocean dynamics. Even so the bias of scientific interest leading up to the expedition seems to have been overridingly biological, fed by the tantalising results brought back by 'natural historians' from a surprising number of earlier research voyages.

In 1831 for example Charles Darwin at the age of 23 had accompanied Captain Fitzroy of H.M. surveying ship *Beagle*, to collect specimens and record the natural history of land and sea during a voyage round the world to close the circuit of meridian distances for the hydrographic office. He had returned with the foundations of a quarter of a century's work which culminated in the publication of *Origin of Species* in 1859. The young T. H. Huxley, appointed in 1846 at the age of 22 as surgeon-naturalist to H.M. surveying frigate *Rattlesnake* had shared Captain Owen Stanley's voyages of exploration to the Great Barrier Reef and New Guinea and collected the floating marine organisms that began his famous classification of the zooplankton.

In addition there was the influential figure of Edward Forbes who had become a professor at both London and Edinburgh universities after spending four years at sea in H.M. survey brig *Beacon* dredging regularly for marine samples in the Aegean. On the basis of this experience, in one restricted area of ocean and with sadly primitive equipment, Forbes had unfortunately concluded in 1843 that there were various depth zones in the sea and that in each of them the number of species of marine life grew fewer with depth until below 300 fathoms there was an 'azoic zone' where neither plants nor animals could live.

In spite of incontestable evidence from dozens of other marine naturalists that Forbes was wrong, his view was an entrenched scientific belief for nearly 30 years. His reputation can perhaps only be redeemed by the suggestion he made four years before his death that two important tasks should be attempted . . . 'one which can be hardly hoped for fulfilment without the help of a steam vessel and continued calm weather, is the dredging of the deeps off the Hebrides in the open ocean', and the other 'though I fear the consummation, however devoutly wished for, is not likely soon to be effected, a series of dredgings between the Zetland and Faroe Isles, where the greatest depth is under 700 fathoms, would throw more light on the natural history of the North Atlantic, and on marine zoology generally, than any investigation that has yet been undertaken.'

When the dynamic Vice-Admiral Richards became Hydrographer of the Royal Navy in 1863 this suggestion by Forbes was put into action and naval survey ships undertook a series of short but immensely valuable scientific cruises. The first voyage was made to the Hebrides in 1868 in H.M.S. *Lightning*. Two leading British naturalists were on board, Dr. Charles Wyville Thomson and Dr. William B. Carpenter. They successfully dredged down to 650 fathoms and brought up numerous representatives of all the known invertebrate groups as well as many species new to science.

In the following year, 1869, three similar voyages were made by H.M.S. *Porcupine* to the west of Iceland, the Bay of Biscay and the Faroe Channel. Wyville Thomson and Carpenter were joined by another scientist, Dr. Gwyn Jeffrys and with improved dredging techniques the three of them managed to bring up bottom fauna from the depth of 2,000 fathoms.

These results were the real spur to the *Challenger* expedition. William Carpenter, from a position of strength as Vice-President of the Royal Society, was able to generate enough interest for it to set up a Circumnavigation Committee with the object of launching a major expedition into 'all three of the world oceans'. What was more, among his fellow members of the famous 'Metaphysical Society' of London, he was able to reach the ear of the Prime Minister, Gladstone and his Chancellor of the Exchequer, Robert Lowe, and persuade them to support the funding of the enterprise.

A great deal of work has recently been done, particularly by the American science-historian Harold L. Burstyn, in uncovering the circumstances which surrounded the *Challenger* expedition. He has pointed out that before the 1870s the established method by which Parliament had kept control of government expenditure had been by voting the absolute minimum to enable the administration to function and in that traditional atmosphere of chronic financial shortage, *Challenger* might very well not have sailed. Fortunately for oceanography Gladstone had become Prime Minister of the new Liberal administration of 1868, after a stint as Chancellor of the Exchequer, in which office he had carried through the reform of the financial structure of the Government by introducing the modern methods of auditing controls, so that money voted by Parliament could be seen to be spent in the way intended.

It could therefore be argued effectively that money appropriated to this scientific venture would be properly accounted for, and with this assurance, and the comment of the popular press of the time that a naval vessel was a formidable consumer of public funds even when she was doing nothing, the money for *Challenger* was found. Even so it was to be a large enough sum to make the expedition the big science project of the day even if the phrase had not been invented. Taking into consideration the length of time involved in planning and fitting, in the voyage itself and in the analysis of the results and the preparation and distribution on an international basis of the resulting report, it was going to take at least 15 years and to cost around £100,000, the equivalent of many millions of pounds in these inflated days. In return for this investment the British Government, without any particular enthusiasm for science as such, considered that it was buying for the country a commanding position in the scientific exploration of the sea to match its undisputed naval supremacy. In this they got a remarkably good bargain as *Challenger's* legend still manages to overshadow and even obscure important contributions made to oceanography by other nations long before she sailed and long after her retirement.

When the Circumnavigation Committee of the Royal Society approached the Admiralty to provide a suitable ship for the expedition it became clear that the ship required would have to be far larger than any normally used for surveying. This ship would have to carry thousands of fathoms of sounding and dredging ropes and provide space for storing an unknown number of marine faunal specimens and water samples, also workrooms, laboratories and accommodation for the extra crew needed to handle the scientific equipment, as well as quarters for the scientists themselves. So the first British ship chosen to be exclusively equipped for oceanographic research was a naval corvette of 1,462 tons displacement. Her spars were reduced, she was stripped of all but two of her guns and she turned her back on a rather undistinguished record of active service in Mexico and Fiji to earn an immortal reputation in the history of science. Her conversion for her task, and in fact the entire business of preparing the expedition including the selection of the commander of the ship, Captain Nares, fell to Vice-Admiral Sir George Richards as Hydrographer of the

Navy. When *Challenger* sailed in December 1872 he said, 'An expedition such as this, which has been the hope and dream of my life, is now on the eve of realisation.' It was only then, when she was safely on her way, that Richards, with the air of a man who had achieved a life's ambition, retired into commercial life and 20 years of cable ships.

The scientists chosen by the Royal Society were led by Wyville Thomson with a young geologist called John Murray, also from Edinburgh, as his personal assistant. Other members of the team were two other naturalists, Moseley and Willemoes-Suhm (who did not survive the voyage), a chemist called Buchanan, and Wild, who, as well as being secretary to Wyville Thomson, was a very good artist. The objects of the expedition were set out in the instructions issued to them by the Admiralty and the Royal Society.

They were to do four things:

> To investigate the physical conditions of the deep sea in the great ocean basins (as far as the great southern ice-barrier) in regard to depth, temperature, circulation, specific gravity and penetration of light.
>
> To determine the chemical composition of the sea water at various depths from the surface to the bottom, the organic matter in solution and the particles in suspension.
>
> To ascertain the physical and chemical character of deep-sea deposits and the sources of these deposits.
>
> To investigate the distribution of organic life at different depths and on the sea floor.

As it turned out these stern instructions were enough to keep all the oceanographers in the world hard at work for a century – not just for the $2\frac{1}{2}$ years of *Challenger's* voyage. Her own response was enough to engage the labour of most of the qualified scientists in Europe and North America for many more years than were taken by the voyage itself. Her greatest successes were in her marine biological collection and the samples brought back of sediments from the sea floor. References to these specimens still figure in the papers of ocean scientists today. A hundred years later historians of science now judge her efforts in marine chemistry, physical oceanography and the study of the circulation of the ocean atmosphere to have been less revolutionary than the sheer volume of her results suggested, partly because of faulty instrument design.

When H.M.S. *Challenger* returned to Britain in May 1876 she had sailed 69,000 nautical miles and made 364 separate stations to take water temperatures and samples, to sound the bottom and trawl and dredge for specimens. Each of these stops involved shortening sail, and lowering into the water twin-bladed propellers driven by a two-cylinder steam engine to allow the ship to navigate accurately and keep her up to the sounding line so that she measured a true vertical distance.

The results of the expedition were, as every oceanographer knows, published in a monumental report of 50 thick volumes, each in its own way the foundation stone of a different area of ocean science today. Yet it was in the publication of this report that the whole *Challenger* venture nearly came unstuck.

The new principles of accountability for public money put the leader of the expedition, Sir Charles Wyville Thomson, at the mercy of the Treasury. Its officials, in the best traditions of niggardly bureaucracy, could hold him personally responsible for every yard of sailcloth and coil of rope that had gone astray throughout the whole voyage. What was worse, when the sorting of specimens, their distribution to the many distinguished scientists who were to examine them, the collection of their findings and the preparation of the report, proved more difficult and therefore more expensive than his original estimate had foreseen, Wyville Thomson was more or less put under threat of public disgrace for extravagance and failure to carry out his trust, because the Treasury refused to renegotiate the contract. Under severe pressure as time for completion of the publication of the report ran out, together with the money, Wyville Thomson broke down, resigned his professorship, and early in 1882, died. Only then, when Wyville Thomson had been lowered into his grave, did the Treasury release funds for the completion of the Report, though still barely enough to allow John Murray to complete the work over another five years. It was a sad sidelight on a great project and one that oceanographers, all over the world, still haggling with governments for the funds they require, are never likely to forget.

The *Challenger* expedition and the preparation and distribution of the data gathered into the Report set some interesting precedents for big science projects today and invite comparison with some aspects of space research, the moon landings and the preliminary exploration of the planets.

First of all, at the time of the *Challenger* expedition, Britain was the

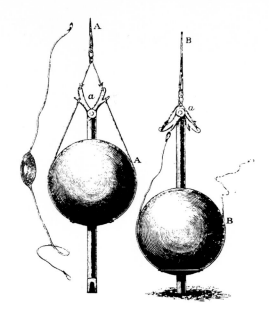

Brooke's Deep Sea Sounding Apparatus of 1845. The cannon ball drives the central sampling tube, originally a quill, into the sediments, releasing itself from the line as it touches the bottom so that the tube alone is raised to the surface.

James Clark Ross with *Erebus* and *Terror* making the first successful deep sounding of 14,550 feet in the South Atlantic in 1840. The ship's boat carrying the line was held in position against wind and current by the oars of a second boat.

H.M.S. *Challenger*, the British naval corvette converted into a research ship to make the first round-the-world oceanographic expedition—December 1872 to May 1876.

An oceanographer makes notes of the time and water depth on the continuous trace drawn by a modern precision echo sounder. (*Photo: National Institute of Oceanography.*)

Echo sounder trace of the sea bed in the Canary Islands just west of Gomera. The deepest point shown is 9,600 feet and the horizontal distance is approximately nine miles. (*Photo: National Institute of Oceanography.*)

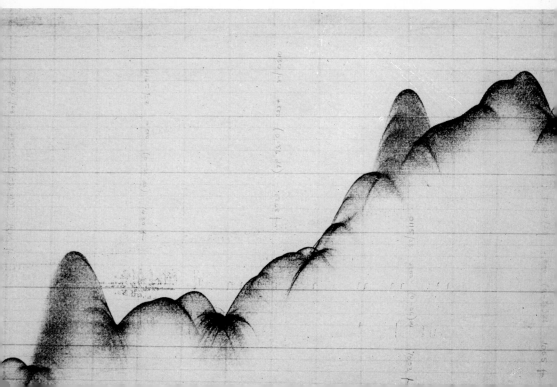

A guyot on dry land? Mount Asmara in Ethiopia has an extraordinary flat top and appears to be built of fragments of volcanic glass. (*Photo: Haroun Tazieff.*)

The central graben of Iceland, several miles across, bounded each side by 100-foot basalt cliffs, may give us an idea of the appearance of the median rift of the Mid-Atlantic Ridge. (*Photo: G. P. L. Walker.*)

A corer fitted with special temperature probes goes over the side of the oceanographic research ship, R.S.S. *Discovery*, to bring back samples of the bottom sediments. (*Photo: National Institute of Oceanography.*)

Haul of rock samples, laboriously dredged from the ocean floor, arrives on deck. (*Photo: National Institute of Oceanography.*)

A geologist in his laboratory at the National Institute of Oceanography, England, measures a section of a long sediment core. (*Photo: National Institute of Oceanography.*)

richest country in the world and therefore the best able to find the necessary funds, yet a hundred years ago the grave social consequences of the industrialisation that had enriched the country had been plainly spelt out to every literate household by the best-selling novels of Charles Dickens. The years in which the financial decisions about *Challenger* were taken were the years in which William Booth was launching the Salvation Army to fight the desperate conditions of the London poor, when an attempt was finally being made to establish compulsory education and to release women and children from what amounted to slave labour in the mines. In modern terms it might seem that only blindness to human suffering could put the doubtful benefits of oceanography ahead of other claims upon the public purse. Oddly enough these claims were never pressed. If they were even considered they were rejected in favour of prestige and future industrial development.

Even on the level of ordinary political tactics, however, it is astounding that Gladstone's administration considered supporting *Challenger*. At the very moment when the financial decision was being taken, the violent Franco-Prussian war had broken out and England was unexpectedly obliged to expand the size of her army to carry out her treaty obligations to her smaller European allies. To find the money the Government had risked its popularity by raising the hated income tax. Yet the money for *Challenger* was found although she was unlikely to come back with anything as directly profitable as the new trade routes and even countries discovered by earlier expeditions.

Secondly, the expedition was mounted in a spirit of scientific internationalism. It could be argued now that Britain could never have managed to protect her huge investment in ocean science or to ensure that the eventual profit returned to her own coffers, but this could be unduly cynical. She made no attempt to keep the benefits of the expedition to herself: the specimens collected were distributed to all the eminent scientists in the field whatever their nationality and at a later stage, the Report was distributed by the Government, at the recommendation of the Royal Society, to suitable institutions all over the world, a process only completed in 1920. It was not difficult to hear the echo of these decisions as the first load of moon rocks was distributed to scientists around the world, or as NASA decided to make its research data freely available to all comers.

Certainly anyone looking for an example to justify the latest 'spendthrift adventure' of pure research will be able to point out the extent to which, as we claim the credit and the prizes of the present great revolutionary period in oceanography, we are still collecting capital gains on investment made in the big science of the past, at a time when the available market information made such investment an act of blind faith.

Looking forward in the 1870s there was no clear road for ocean research to follow – only a number of winding paths leading into different parts of an untrodden thicket. Looking back from the 1970s we can see very well which path was really the first section of a broad highway of discovery. Though most of the hundreds of routine round-the-world soundings made by *Challenger* probed nothing as sensational as Pacific trenches or Atlantic mountains, they were at least a prominent signpost in the right direction. In the decades that followed, soundings accumulated from survey ships and cable ships as the laying of telegraph lines reached a climax all over the world and as more and more research ships were launched by all the maritime nations anxious to stay in the race for whatever rewards – prestige or profit – the study of the oceans might bring. For years after *Challenger* had left the scene, many historic vessels – the U.S.S. *Albatross*, the U.S. Coast Survey steamer *Blake*, the French ships *Travailleur* and *Talisman*, the Russian *Vitiaz*, the German *Valdivia* – were roaming the seas and building up a great bank of skill and information. As a result it became obvious by the end of the nineteenth century that a comprehensive ocean chart was required to plot all the new soundings that had been collected.

Making charts for navigational purposes had been a major undertaking of all seafaring nations for more than a hundred years. By the time a Hydrographer was appointed to the Royal Navy in 1795 more British ships were being wrecked for lack of charts than were being lost through enemy action. This desperate state of affairs, together with the growing strategic and commercial importance of accurate navigation, amply justified the great expense and effort of surveying. The charting of the deep ocean, however, was clearly beyond the resources of any single nation. It had to be a co-operative undertaking, and, by a strange quirk of oceanographic history, the task of co-ordinating the international effort fell to one of the least pugnacious and most diminutive of states, the Principality of Monaco.

The great chart project was first discussed in 1899 by a committee of eminent scientists at the VIIth International Geographical Congress in

Berlin. In 1903 a second meeting was held in Wiesbaden under the chairmanship of Prince Albert I of Monaco, a dedicated oceanographer who had made particularly accurate soundings from his private research ships with a machine of his own design. After this meeting the Prince's own 'scientific office' in Monaco undertook the production of the first 'General Bathymetric Chart of the Oceans' – the 24 sheets usually called the 'GEBCO' or 'Monaco' Chart.

This first chart, with 18,400 deep soundings plotted, was produced at incredible speed and submitted for approval to the VIIIth Congress held in Washington the very next year, 1904. Since that time the chart has been updated with new soundings every few years but after the death of Prince Albert in 1922 the task was taken up by the then newly-formed International Hydrographic Bureau, whose headquarters, at the Prince's invitation, had been set up in sunny Monaco.

Since 1965 the work of preparing new five-yearly editions has been shared out among 19 member nations. Even Russia, who is really only an observer and not a member of the Bureau, looks after four of the sheets. The reason for this division of labour is simple: even by the time the third edition was started in 1932 the number of deep soundings plotted on 1,001 preliminary sheets at a 1/1,000,000 scale had reached 370,000, of which 54,518 were used in the final 1/10,000,000 version.

Since ocean science has really begun to leap ahead in the last decade a certain amount of quiet muttering has been heard that the GEBCO sheets are still not keeping pace with oceanographers' requirements, either in style of presentation or in speed of response to new data. For serious research work the kind of charts now required are the 40 sheets covering the Pacific Ocean alone, recently published by a team from the Scripps Institution of Oceanography which cannot fail to give more accurate contours than the 24 GEBCO sheets for the whole world.

One of the most refreshing and delightful characteristics of oceanography at this moment in time is the broad and simple scale on which its major issues can be described and discussed. While so many other sciences are virtually impenetrable to the non-specialist, so deeply are they buried under layers of minutiae and mathematics, oceanography is still concerned with building up a general picture of what is going on in the sea, and this, instead of being a weakness, has turned out to be its strength. The most exciting and intellectually challenging advances of the last decade, however limited or detailed the observations and measurements on which they were based, have been changes in the view of the ocean as a whole, indeed of the behaviour and structure of the planet.

For a modern oceanographer and certainly for a marine geologist the graphic summary of what has been discovered on the ocean floor is extraordinarily important. It can even be argued that until a certain critical amount of information had been organised, as it is on a chart, into an instantly intelligible communication, independent of language, the real interpretative breakthrough could not be achieved.

3

Putting it on the Map

IN THE TACTLESS WAY THAT SCIENCE HAS OF PROFITING FROM THE DISASTER of war to make some indisputable advance in human achievement, oceanography made a great leap forward after World War I with the development of the echo sounder. It came into the hands of oceanographers as a peacetime adaptation of a weapon developed by the collaboration of an Englishman, Ernest Rutherford and a Frenchman, Pierre Langevin, to enable allied destroyers and Q–ships to hunt German U–boats.

Later on their efforts were to flower into the well-known ASDIC equipment (its title is the acronym of the Allied Submarine Detection Investigating Committee), which transmits a burst of vibrations from a quartz oscillator. When this vibration meets a reflecting surface it is returned as an 'echo' which can be picked up by a receiver. The travel time through the water of the vibration between transmission and reception reveals the distance of the reflector. If a reflection is found where no reflection should be there is a very good chance that it is a submarine.

The adaptation of this principle to oceanographic work by bouncing echoes off the sea floor was one of the first technical achievements of the new British Admiralty Research Laboratories established in 1921 close beside the National Physical Laboratory at Teddington. The credit for the successful design work belongs to B. S. Smith of the Acoustics Group who produced the first telephone audio-frequency system for obtaining depth. The sound source of the equipment was a steel diaphragm five or six inches in diameter, which emitted a train of audio-frequency waves (in other words a loud noise) when struck regular blows from an electro-magnetically operated hammer. A small hydrophone picked up the returning echo from the sea bed.

The operator wearing headphones turned a calibrated dial until he could hear the returning echo and he could then read off from the dial the time taken by the signal to reach the bottom and return. This was then translated into fathoms or feet and corrected for the velocity of sound in sea water at various temperatures. The depth reading had to be recorded by hand for subsequent plotting on a chart. It was all quite a performance and it could give the operator a nasty headache but it was a considerable improvement on several hours spent struggling with five miles of wet steel wire and a winch.

After sea trials in 1923 the echo sounder was handed over by the Admiralty to Henry Hughes & Sons (later to become Kelvin & Hughes Ltd.) for commercial manufacture and before long ships of all nations had versions of the new acoustic instrument and results began to appear. With an advanced electronic echo sounder a modern ship can steam along her chosen track while a continuous profile of the sea bottom is recorded on a moving roll of paper with a chemically treated surface.

Up until 1934, however, most surveying ships were using equipment that was really only satisfactory at depths of around 200 fathoms which practically confined them to working over the continental shelf. The limitation was the basic problem of transmitting sound in water. Sound is 'attenuated' or lost per unit distance travelled in water, approximately in proportion to the square of its frequency, so if the returning echo is to be heard at all the original source must be of very low frequency. Unfortunately a low frequency source is virtually non-directional so it is very difficult to tell whether the reflecting surface is on the ship's track or somewhere off to one side and the deeper the water the worse the problem becomes.

So the Admiralty Acoustics Section took the well-trodden path back to the drawing board and in 1930 the old nerve-racking procedures of hammer and earphones was overtaken by progress. Three scientists, A. B. Wood, F. D. Smith and J. A. McGeachy, succeeded in making a powerful high-frequency directional version of the echo sounder using a condenser discharge instead of a hammer and also using a new 'magneto-striction' transmitter and receiver.

Magneto-striction is a phenomenon in which a ferromagnetic material, for example nickel, changes its linear dimensions when magnetised or, when its dimensions are forcibly changed, shows a change in its magnetisation, which makes it suitable for use as an 'oscillator' or 'transducer'. This equipment sent out a narrow enough beam for the position of the reflecting surface to be properly located and so make deep sounding really simple. The signal was recorded by a stylus passing a current onto a chemically coated paper surface to expose a dark line, a technique inherited direct from Michael Faraday a hundred years earlier. At long last the lead and line inherited from the days of Herodotus were finally laid aside.

Even before the magneto-striction echo sounder for deep work became commercially available, enthusiastic oceanographers all over the world had managed to get their hands on some kind of sonic sounding equipment. The Danish vessel *Dana*, exploring in the Indian Ocean between 1920 and 1922, found an underwater rise that she named the 'Carlsberg Ridge' after the lager company that had contributed funds to the expedition.

Between 1925 and 1927 the German research ship *Meteor* made detailed explorations in the North Atlantic. She made a number of traverses over the old 'Telegraphic Plateau' whose existence *Challenger* had confirmed by lead and line in 1873, and found that it was really a rugged range of submerged mountains that came to be known as the Mid-Atlantic Ridge. Then in 1929 the American oceanographic vessel *Carnegie* used echo sounding over the Albatross Rise, discovered long before in the Pacific by Louis Agassiz, and found a substantial oceanic mountain range. In the light of what was to come a quarter of a century later, these successful soundings of submarine mountains were key discoveries for geophysics, though at the time they were unrelated and inexplicable curiosities.

Even when more reliable deep echo sounders were developed such lucky finds were rare. Marine geologists had not reached the point of being able to formulate any general theory about the sea floor to guide research and in any case, although navigational echo sounders of the shallow-water kind were being fitted in all survey ships and even some merchantmen, the deep-water equipment was expensive enough to be confined to a few specialised research ships. One of them, nevertheless, an

18 kV model, found its way into a warship, the U.S.S. *Cape Johnson*, operating in the Pacific during World War II. It suggests more than a happy coincidence that such an echo sounder was available when, for a few months in 1945, the commander of the ship, a reserve lieutenant, was a Princeton geologist, Dr. Harry H. Hess, who had already distinguished himself by producing the best available bathymetric chart of the North Pacific. The *Cape Johnson* took part in no less than five major wartime

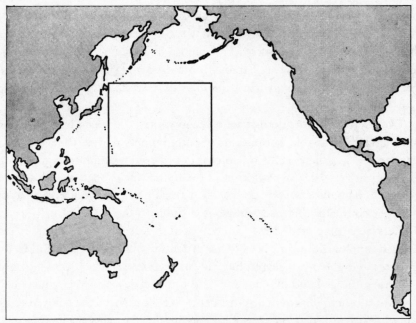

AREA IN THE NORTH-WEST PACIFIC IN WHICH GUYOTS WERE FIRST IDENTIFIED.

landings but in less martial moments she managed to give oceanography a push into the future.

　　Professor Hess is now dead and the facts of his wartime adventures in surreptitious science are rapidly fading into legend. Each time the story is retold the rank of the commander gets lower and the size of the ship gets smaller, declining from a fleet oiler to an amphibious landing craft. As it may only be a few more years before she is diminished into a footloose longboat in the charge of a midshipman it is worth recording that she was in fact a freighter converted into a troop transport of 5,668 tons displacement and that Hess ended his career as a U.S.N. Reserve Rear-Admiral, none of which makes the story any less intriguing.

Hess could not bear the idea of all the good ship-time that was going to waste, oceanographically speaking, as his ship moved around the North-West Pacific and, in spite of the risks from enemy submarines capable of picking up the energy pulses, he persuaded his crew, largely of wartime volunteers, to operate the echo sounder whenever possible. In the course of several million square miles the recorder drew the profiles of about 20 'curious flat-topped peaks' which looked exactly like truncated volcanic islands except that they were all somewhere between 3,000 and 6,000 feet below the surface. Some were single, others stood in groups, their flat tops all at the same elevation, suggesting that they were platforms cut by waves when the peaks stood above surface level. Just to confuse matters, some of them, though close together, differed in height by as much as 1,000 feet, so

SHAPE OF A TYPICAL GUYOT BASED ON ECHO SOUNDER TRACE

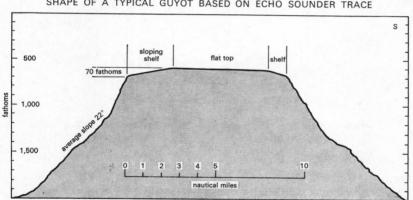

they could not all have been worn down at the same sea level and they must have suffered at different dates. Hess first discovered the mysterious seamounts in 1944 and named then 'guyots' in honour of a distinguished Swiss Professor of Geology at Princeton, Arnold Guyot, to whose chair he himself succeeded. He spent a lot of time pondering on their origin and significance. From his own ship's records and the files of the U.S.N. Hydrographic Office he found another 140 guyots and now almost 500 of them have been found in the Pacific as well as a few in the Atlantic. Two questions about them needed answering: first the dates at which their flat 'wave-planed' upper surfaces had been formed; and second, their relation-ship to the coral atolls scattered among the islands in the same general area.

Atolls are in some ways as odd as guyots. Seen from the air or better still, from space, they are circlets of coral wholly or partially enclosing a lagoon. Some of them are little more than a strip of coral reef but others are wide enough and sufficiently covered by debris of some kind or another to support vegetation, even trees. On his celebrated voyage in the *Beagle* between 1831 and 1836, without benefit of echo sounding, Charles Darwin deduced from the biology of the coral polyp the proposal that atolls were coral crowns built upon the tops of sinking volcanoes. He found that reef corals can only live in the sunlit upper levels of the sea, roughly the top 200 feet and that they build their skeletal structures of calcium carbonate one above the other. He argued that to remain always at sea level the reef building corals must be keeping pace with a slowly sinking platform.

Even in an area so extraordinarily rich in marine volcanoes this was a bold idea but a theory of enormous vertical movements of the earth's crust

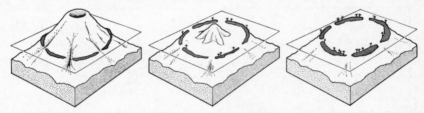

Three stages in the formation of an atoll as a coral reef builds upwards on a sinking volcano.

was fashionable at the time. As Darwin himself had recently seen such convincing signs of them as wave-cut beaches uplifted 14,000 feet above sea level in the high mountains of Patagonia, why should some downward movements not take place as well? As it happened the state of engineering prevailing in 1836 made the suggestion just about uncheckable, although it was never forgotten. Other theories of atoll formation came and went, deflated by more obvious objections than any that could be levelled at Darwin's logical conclusion.

Several attempts were made between 1896 and 1900 to see if Darwin was right. The Royal Society had set up a Coral Reef Committee which organised some deep borings into the coral of Funafuti Atoll in the Ellice Islands and in one attempt they reached a depth of 114 feet. The cores, when they were analysed, were found to consist entirely of dead coral of the shallow water type but as no one could be sure that the coral brought up was in its original position of growth and had not fallen down as debris from the living top of the reef no conclusion could be drawn. Darwin's vindication had to wait for an age of higher technology and more money.

It was not until 1950–51 that two Cambridge oceanographers, Dr. Tom Gaskell and Dr. John Swallow, aboard a British hydrographic survey ship, a later H.M.S. *Challenger*, visited Funafuti Atoll in the course of a world cruise. Using seismic techniques they established that under about 1,800 feet of coral limestone below the lagoon there was a large hump of light volcanic rock rising from a deeper harder volcanic core. Later still, U.S. Army engineers drilling on Eniwetok Atoll in connection with the atomic bomb tests sank a shaft down through more than 8,000 feet of coral before they hit volcanic rock.

Writing a paper about his guyots just after World War II, Hess didn't have the advantage of either of these observations. All he knew was that if guyots were going to be included with atolls in the category of drowned ancient volcanic islands they created a wholly new enigma. Why were there no coral crowns on the submerged tops of the guyots? Had they sunk too rapidly for corals to keep pace with them? Yet the flat tops of many guyots were so broad in diameter that they must have spent a long period at the surface for the waves to do their destructive work. Suppose the temperature of the sea had changed so that corals could not grow; had the sea level been dramatically lower in a long period of glaciation? Had guyots indeed subsided so long ago that the reef building coral organisms did not even exist? True to form, the new research tool, the echo sounder, was raising as many problems as it solved.

Having digressed so far in pursuit of guyots, largely because they led their discoverer, Harry H. Hess, on to make theoretical deductions about them that were of the utmost importance to all current oceanographical work, it seems a pity to abandon them without adding the newest clue to the mystery of their origins.

Between 1967 and 1969 a team of Italian, French and American scientists led by the vulcanologist Haroun Tazieff braved hostile tribesmen and soaring temperatures in three expeditions to survey the Afar Triangle at the south-west corner of the Red Sea and study the complexities of the northern stretch of the rift valley system of East Africa. They found what may be the one area on earth that is a stretch of true ocean floor temporarily raised above sea level. Within it were exposed on land flat-topped mountains that look remarkably like guyots. One, Mount Asmara, 1,200 feet high, tapers from a base a mile and a quarter across to a top of two-thirds of a mile. The team's geologists have reported that the whole mountain seems to have been made of successive layers of tiny glass fragments called Hyaloclastites. Tazieff, reporting on the expedition, says that he has observed these glassy fragments formed when hot lava is produced in a violent underwater eruption. Apparently the primary eruption can tear the molten magma into pieces and fling them into the

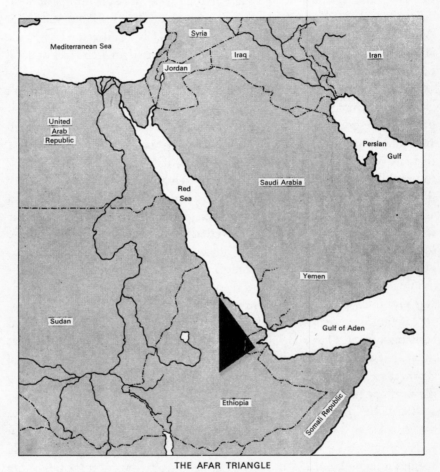

THE AFAR TRIANGLE

This land area seems to have started as part of the floor of the Red Sea which, together with the Gulf of Aden, is probably a young ocean.

water which, when it touches them, turns into superheated steam. This in turn causes explosions that turn the lumps of lava into glass fragments that may be thrown as high as a kilometre above sea level. When they fall back into the water they form very regular rims of ash round the vent of the volcano. Mount Asmara seems to have been built up in this way by successive eruptions and depositions of glass, without ever rising above the surface of the water that originally covered the whole area.

If Tazieff's deductions about Mount Asmara are correct they may show how at least some guyots were built, also why their flat tops are at different levels. Only a particularly active volcano would manage, either in one or in several successive eruptions, to break the water surface and create an ash and lava cone that would be unaffected by the secondary explosions due to water. This explanation might cover the initial stage of formation of atolls as well, the difference between a flat top and a clearly defined rim perhaps depending on the size, shape and behaviour of the volcanic vent, possibly even on the depth of the eruption below the surface. The Afar Triangle has in fact revealed at least one old and eroded ash ring overlain with corals and marine animals of the Pleistocene period (around two million years ago).

The steady growth of the coral reefs still demands that the Pacific atolls should be sinking, which leads us straight into one of the chicken-and-egg paradoxes of science. If Hess had known of the existence on land of Mount Asmara he might have missed the idea that guyots had sunk with the sea floor and his later, dramatically influential theories based on this notion might never have been born. Then there might have been no expedition to look with such sharp eyes at the Afar Triangle because of its association with a rift system.

As it was, after World War II, guyots simply got added to the slowly developing deep-sea chart, and research ships, in a pleasantly irresponsible state of ignorance, went on enjoying themselves with their new echo sounders. One of the more light-hearted results was the speeding up of a curious 'depth competition' that had quietly started in the nineteenth century to sound the deepest part of the ocean floor. The fact that nobody admitted it was happening did not make it any less of a contest. As we have seen, the first H.M.S. *Challenger* had found 4,500 fathoms (27,000 feet) in the Marianas Trench in 1873. Almost at once the U.S.S. *Tuscarora* reached 4,655 fathoms (27,930 feet) in the Kuril Trench and in 1895 *Penguin* reached 5,155 fathoms (30,930 feet) in the Kermadec Trench, the first sounding below 5,000 fathoms.

By ship after ship, a few fathoms at a time, the record was pushed down. The German ship *Planet* held it in 1912 with a line sounding in the Philippines Trench until the U.S.S. *Ramapo* found 5,673 fathoms in the Japan Trench and the German ship *Emden* claimed 5,850 fathoms, again in

Profiles across two Pacific trenches drawn from precision
depth recordings with a vertical exaggeration of 22:1.

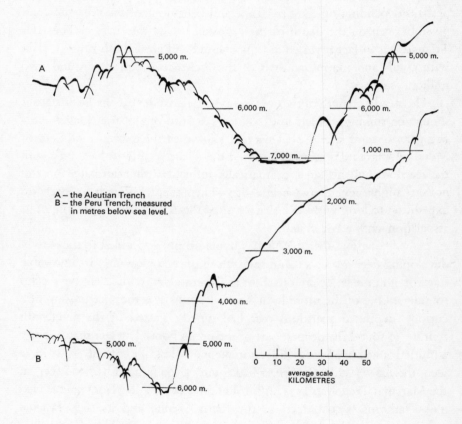

A – the Aleutian Trench
B – the Peru Trench, measured
   in metres below sea level.

the Philippines Trench, in 1927. The Dutch ship *Willebord Snellius* went to look for the Emden Deep in 1930 but at the position given found nothing deeper than 5,509 fathoms. The *Emden*'s record tottered. Then in 1944 the celebrated U.S.S. *Cape Johnson* weighed in and after dismissing the Emden Deep as a mere 5,492 fathoms, came up with a new 5,686 fathom deep of her own, the Cape Johnson Deep in the Mindinao Trench. Finally, at what may be the close of play, the new H.M.S. *Challenger* in 1950 sounded 5,940 fathoms (35,640 feet) in the Marianas Trench, a record that at the time of writing has not been broken.

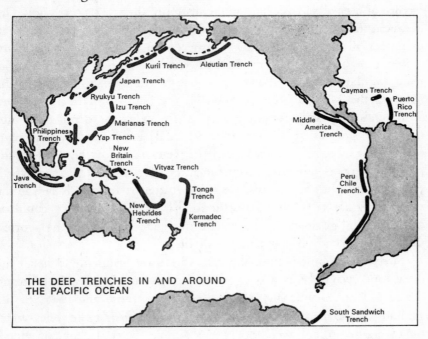

THE DEEP TRENCHES IN AND AROUND
THE PACIFIC OCEAN

The big important fact that emerged from it all was that the great oceanic trenches where all these deeps were plumbed were all very similar in general configuration and within a few thousand feet all of much the same depth. To the record-breakers who discovered this it may have been a disappointment. Their preoccupations were with achieving the exciting profiles promised by these staggering but inexplicable physical scars in the ocean floor, and with the problems of coaxing their echo sounders into trying to attain the sharpness of detail that is still waiting for a higher technology of ultra-narrow 'beams' of energy. It is very doubtful if anyone involved ever guessed at the importance trenches would have today as key structures in the geophysicists' systematic view of the world.

The echo soundings have revealed some other unexpected deep-ocean features in recent years: the great fracture zones or faults that split the ocean floor for extraordinarily long distances, especially in the Pacific. These splits first showed up on charts as significantly large differences in depth and their discovery was largely due to the curiosity of Professor H. W. Menard and a research team at Scripps. From 1944 onwards they began slowly piecing together soundings made off Northern California by the Institution's own research ships until they built up a disconnected picture of a possible fault line with cliffs more than 5,000 feet high. The dramatic topography of the faults only emerged bit by bit as ship-time became available for proper zig-zag survey tracks to be run over them.

Between 1950 and 1954, 13 big faults were found in the Eastern Pacific, many of them now well known as the Mendocino, the Pioneer, the Murray, the Molokai, the Clarion and the Clipperton fracture zones. Sometimes they are astounding chasms that crack the ocean floor into troughs as much as 200 kilometres wide and deeper than any other points of the sea except the trenches. Some of them have now been traced for more than 3,000 kilometres, stretching from the western coast of North America until they lose themselves in the scatter of volcanic islands in the west. They are so straight for so long that at first sight they could be mistaken for 'great circle' tracks, the navigator's lines of the shortest distance between two points on the globe.

Similar fractures, though shorter, were soon being found in all the oceans and even today new ones or new stretches of old ones are still being discovered as research ships gradually traverse more and more of the ground. Like so many other discoveries in the sea, the fracture zones, when they were first found, were totally inexplicable. The only hope of understanding them was to struggle on until this initial 'descriptive' phase of the science established so decisively by the echo sounder yielded enough material for people to start to construct some kind of overall theoretical framework. More than a decade later this would carry everyone forward into oceanography's 'dynamic' phase when it would be possible to see how all the different bits fitted together into a working system.

Echo sounding was not the only technique available to oceanographers attempting to decipher the structure and history of the ocean basins. Several other approaches were tried, some of the most successful also owing an unexpected debt to the chances of war.

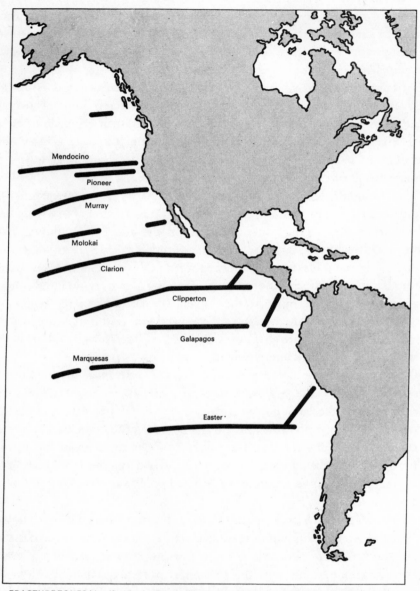

Mendocino

Pioneer

Murray

Molokai

Clarion

Clipperton

Galapagos

Marquesas

Easter ·

FRACTURE ZONES identified in the Eastern Pacific by 1964, before their full westward extension had been traced.

Because they were neutrals and could neither fight nor go to sea, a group of oceanographers in Sweden between 1939 and 1945 occupied their frustrated talents in planning and raising the money for the first Swedish round-the-world expedition to explore the deep ocean: what eventually became the voyage of the *Albatross* in 1947–48.

Starting from the simple notion that the fossil record in the sedimentary rocks on land was badly distorted, both by millions of years of erosion and earth movements that displaced the layers from their original position, the Swedish scientists wanted to bring back from the deep ocean floor samples of sediment as it had been laid down, to examine at least the micro-fossils of the sea in their original condition.

A straightforward sample of material from the sea floor, examined under a microscope, will yield a great deal of information about what has been happening in the water and even in the atmosphere above. The bottom is the final resting place of all indestructible debris, from muds and clays, airborne dusts, fragments brought into the atmosphere from space, ashes from volcanic eruptions, often spreading out over hundreds, even thousands, of square miles, to the remains of marine creatures, especially the shells of the minute creatures of the plankton. Within the category of sedimentology there are all kinds of highly specialised sub-disciplines dedicated to analysing and interpreting this material.

Rates of sedimentation vary all over the ocean. Near the coasts they can be high because large quantities of material are brought down by erosion from the land and spread out by rivers. Water loaded with this material in suspension can move through clear water as a fast 'turbidity current' and travel for very long distances over the bottom. Sediments transported by the currents can be spread so evenly over the bottom that for hundreds of miles at a time the level may only vary by a foot or two in a mile.

Far away from land the rate of sedimentation may be much lower depending on the amount of marine life in the water. Some parts of the sea are almost biological deserts, because nutrients are not available to sustain the planktonic plants and the animals that feed on them. And in the deepest water only the animal shells built of silica are deposited because the calcium-based shells are dissolved by seawater at a certain pressure and temperature level and do not ever reach the bottom. Sometimes the overall rate of sediment deposition may be only a centimetre or two every thousand years.

If samples can be brought up as cores, preserving the sediments in their layered sequence, the analysis can be applied to a detailed interpretation of the state of the ocean in the past, and the deeper the sampler digs the further back into the past it will go. Cores between six and ten feet long were obtained in the early thirties by a rather dangerous explosive corer that discharged itself like a submarine gun when it came in contact with the bottom. The Swedes decided to improve on this and go deeper. They developed a design made by Dr. B. Kullenberg, using a heavy steel coring tube which was dropped into the sediment layers using the high water pressure of the deep ocean to enable a piston to retain a column of sediment inside the tube when it was withdrawn from the sea bed. This is effectively the modern vacuum core sampler and it can mine cores of 70 feet long.

Today practically every oceanographic ship carries vacuum corers as standard equipment to whatever interesting patch of ocean she may be bound. The cores, eased tenderly from their piston tube onto the ship's deck, are carefully stored in chilled lockers until they reach the refrigerated 'core libraries' of the major oceanographic research institutions where they are kept at the near-freezing temperature of the deep sea floor.

At the slowest rates of sedimentation from such places as the Central Pacific, such a core could cover a time span of between 20 and 30 million years. Even at faster rates of deposition it could go back three million years, relying for greater age on finding places where faulting of some kind has exposed older deposits.

Detailed analysis of cores, especially of the microscopic biological debris, has gradually produced a body of knowledge comparable to the fossil record compiled by palaeontologists from rocks on land and a time-scale by which all cores can be comparatively dated.

To reach back further than this into the history of the oceans is beyond the capacity of an ordinary research ship and requires drilling. It was the chance of drilling through all the sediment layers laid down in the Pacific since the ocean was formed (and as it was thought to be 'primeval' that meant almost to the earliest days of the earth's history) that made marine geologists and palaeontologists line up so enthusiastically behind the Mohole idea, the proposal to drill down to the rocks of the mantle, down at least to the Mohorovičić discontinuity where the mantle might be said truly to begin.

At the time of writing the American drilling ship *Glomar Challenger*, the successor to the ill-fated Mohole venture, is embarking on a 30-month extension of the first successful programme of deep drilling in selected oceanic sites to try to retrieve cores that date right back to the underlying basement rock. Unfortunately it turns out that the sediment record is not so easy to get hold of after all. At the very places where the sediments should be deepest and oldest and most revealing there are layers in which the silica-based skeletons of the tiny radiolarian animals seem to have suffered some chemical change and turned into 'chert', a hard material rather like flint that effectively stops the drill by wearing out the bit. On her first voyages the ship did not have a facility for changing the bit and replacing the drill in the same hole, so a thick layer of chert was a barrier to the sediment layers that may lie below it.

At the time the *Albatross* was making her circumnavigation, deep ocean drilling was not even a remote possibility. Even coring, now that the echo sounder was providing sea floor profiles while the ship steamed along, meant a special and deliberate effort that stopped the ship for hours at a time. So to enlarge the area of information that could be investigated in dealing with sediments the ingenious Swedes developed another advanced technique. Professor W. Weibull, an explosives expert from the Bofors armament works, adapted a method used in mineral and oil prospecting on land, and started exploding depth charges deep down in the sea, picking up by hydrophone the echoes reflected by the top and bottom surfaces of the sediment carpet.

The method was soon improved and developed into a highly effective way of mapping the sediment cover over large areas, especially by Dr. Maurice Ewing and his brother John and a whole group of indefatigable sediment hunters from the Lamont-Doherty Geological Observatory of Columbia University. Today, after a good deal of improvisation and development, a research ship is likely to use a compressed air gun to make a steady train of explosions and lay out a floating line of geophones behind her. This reflection technique requires complicated equipment to sort out all the different sound waves when they are picked up but it gives a very detailed picture of the sediment layers. Unless the velocity of the sound waves in each layer is established by some other method, however, reflection does not provide an accurate account of the depth of the layers. That 'other method' is usually the older and cruder 'refraction'. This also uses an explosion of some kind for its sound source but it is an easier job to sort out the different waves as the technique uses two separate stations –

one the ship and the other either a second ship or a moored buoy carrying instruments – to pick up and to compare the direct sound and the sound refracted by the various layers, even layers 10 miles below the bottom. In this way, by using miniature man-made earthquake shocks, the 'velocity' of each layer can be established and its composition inferred.

Exactly what the rock layers below the sediments consist of is still not properly understood, though the travel waves of the sound waves reveal some of their characteristics. What was discovered is that the rocks under

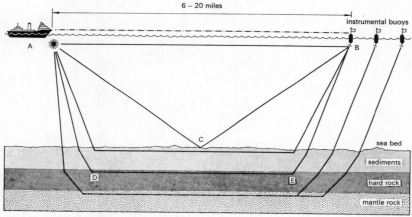

The method of investigating sediment and rock layers by sound waves from a surface explosion. Dotted line represents radio waves between buoys and ship. AB is direct sound wave through water. ACB is sound wave *reflected* by sea bed layer. ADEB is sound wave refracted by rock layers.

the sea floor are quite different from continental rocks both in density and thickness. Where the continental rocks are roughly 20 miles thick and relatively light, consisting of acidic, granitic volcanic material, the sea floor is far more dense and only about four or five miles thick, consisting of basic rocks and a top layer of lavas, and this distinction, as we shall see, is of fundamental importance to geophysics.

Refraction had been put to work, mainly by Maurice Ewing, before World War II but in those far-off days both the explosive charge and the detecting instruments were lowered to the sea bed, generally the bed of the continental shelf, though attempts were made to get results from the deep ocean floor. It was only in 1939, just before the war began, that Sir Edward Bullard made the first trial shot in the English Channel with both the explosion and the instruments at the surface of the water, where, much more conveniently, they are used today. It was nevertheless the results of primitive pre-war refraction shooting that established the astonishing five to ten

mile thickness of the sediments of the continental shelves. It was also established after the war, when the refraction technique improved, that the
sediment layers over the oceans, wherever they were investigated, were
far thinner than they ought to have been if the sea was truly primeval.
The missing sediments were in fact one of the major clues that the oceans
were relatively young – at most only as old as five per cent of geological
time.

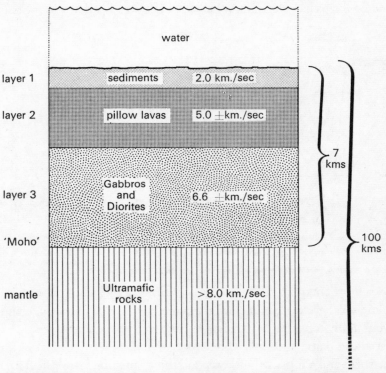

The oceanic crustal layers showing the approximate velocity of P waves in kilometres per second.
The Mohorovičić discontinuity is treated as a layer boundary.

In the end, however, it was not the miniature man-made earthquakes
caused by depth charges but the full-scale and alarming natural ones that
brought about what may well be the most important single discovery not
only in the short history of oceanography but in the study of the earth
itself.

The devastation and loss of life caused by the relatively frequent earthquake disasters have managed, for several hundred years now, to break
down both national barriers to co-operation and the penny-pinching
policies of individual governments. Even if prevention of earthquakes was

beyond all human power, research into their cause and possible prediction were politically and economically desirable. So for a long time money has been available for the establishment of chains of seismological stations for earthquake detection, to pinpoint the earthquake epicentres, the position from which the detectable vibrations radiate outwards through the earth. Latterly, since it became important to be able to distinguish between natural shocks and man-made ones from nuclear weapons, huge and sensitive seismic arrays have been set up backed by computers, and seismology has become an important discipline, a big science in its own right.

In 1949 two scientists from the California Institute of Technology, Gutenberg and Richter, published a series of maps that plotted all the earthquake epicentres recorded in the world between the years 1922 and 1944. They made a sketchy pattern of dots that crossed continents, skirted island chains and in places ran clear through the middle of oceans. In particular they confirmed some earlier neglected results from a seismologist called Tams who by 1927 had found that most of the earthquakes in the Atlantic could be plotted along a line that coincided with the newly located Mid-Atlantic Ridge.

The new Gutenberg and Richter maps were immediately interesting to Maurice Ewing and his team at Lamont. Quite apart from the normal geophysical interest in anything new about the earth that might be turned up by the seismologists, they had themselves visited the Mid-Atlantic Ridge only the previous summer in a preliminary exploration of its reflecting surfaces with their own explosive shocks.

In 1953 they went back to the Atlantic in the Observatory's own research ship *Vema* and made a number of traverses over the ridge. The results of the work were analysed and drawn up as a partial elevation, a 'physiographic diagram', by another team member, Marie Tharp, and in the course of working up a preliminary sketch of the relief she discovered that she was drawing a deep rift valley splitting the crest of the mountains in every section of the range. Checked and rechecked on the profile records it took shape as a valley of sensational dimensions. It averaged 6,000 feet in depth and its width ranged between eight and thirty miles over hundreds of miles of length: in comparison the Grand Canyon of the Colorado River is only 4,000 feet deep and never more than 18 miles wide over a total length of 60 miles.

As it happened, the Lamont observations were paralleled by another expedition to the Mid-Atlantic Ridge that year by the British ship H.M.S.

Challenger, modern namesake of the famous nineteenth century explorer. In the course of a number of traverses between 46° 30′ N. and 47° 45′ N. the Cambridge oceanographer Maurice Hill had also found that the one clear feature of the broken rocky topography of the ridge was the same deep central valley running from north to south between the lines of high peaks. He published news of the valley he had found without drawing any conclusions about it but the Lamont team were sufficiently excited by their discovery to be looking in all directions for clues to its significance.

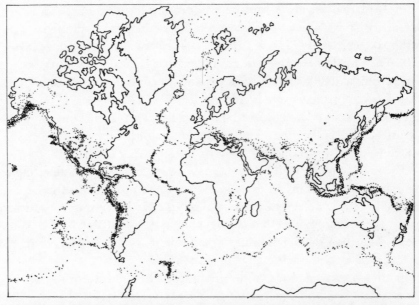

Epicentres of earthquakes throughout the world between 1961-67. Earlier plots of this kind in the 1940s and 1950s led to the discovery of the mid-ocean ridge system (based on plot issued by U.S. Coast and Geodetic Survey).

While they were still drawing their diagram and brooding over the results, advances were also being made by seismology. New recording stations set up in South Africa enabled the French seismologist Rothé to fill in some of the gaps in the earthquake map and definitely link the line of mid-Atlantic earthquakes to an equally well-defined line dividing the Indian Ocean, nosing its way into the Gulf of Aden and the Red Sea and curling down the great rift valleys of East Africa. There were signs of a continuation of the line as a wide loop through the Southern Ocean round Antarctica and it was already clear that the epicentres could be plotted through the Arctic basin to the mouth of the Lena River in Siberia. All these earthquakes were relatively shallow; their epicentres located roughly at a depth of around 30 kilometres below the earth's surface, notably different from the deep earthquakes of the Pacific 'ring of fire' where epicentres of up to 700 kilometres deep were all associated with the great ocean trenches, and with either the chains or arcs of volcanic islands, or with great continental volcanic ranges. A similar broad band of land

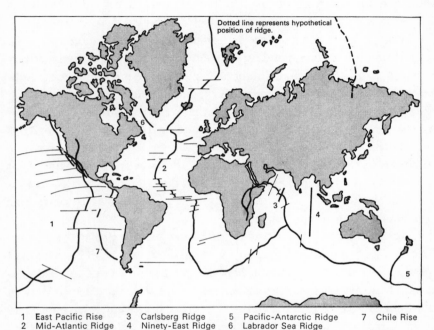

Dotted line represents hypothetical position of ridge.

| 1 | East Pacific Rise | 3 | Carlsberg Ridge | 5 | Pacific-Antarctic Ridge | 7 | Chile Rise |
| 2 | Mid-Atlantic Ridge | 4 | Ninety-East Ridge | 6 | Labrador Sea Ridge | | |

By the early 1960s the outline of the world-wide mid-ocean ridge system could be plotted on a map.

earthquakes (though their epicentres were all less than half as deep), ran through the Himalayas and the high mountains of Persia, Turkey and the Mediterranean into the Alps.

At Lamont, Maurice Ewing and his associate Bruce Heezen studied the earthquake map and Marie Tharp's profiles of the Mid-Atlantic Ridge and in 1956 they staked their reputations on a daring prediction. They said that wherever the line of shallow epicentres crossed the ocean there would also be found ranges of mid-ocean mountains that would form a more or less continuous rifted ridge system around the world.

They were not kept long in suspense before new evidence came in to support their bold prophecy. The year 1957-58 was designated an International Geophysical Year and practically every sea-worthy research ship in the world was afloat, taking part in a co-ordinated programme of observations. Steaming across the earthquake line in search of the predicted mountains was high on the list of interesting things to do and successful profiles of the phenomenon were produced by echo sounders from many different parts of the ocean. *Vema* herself set out to traverse the seismic belt in the Indian Ocean south-west of the Carlsberg Ridge and, sure enough, there was a new stretch of rifted mountain range. She went on to find a further stretch of it along the epicentre belt south-east of New Zealand.

Another expedition from the Scripps Institution of Oceanography even found some of it corresponding with an odd stretch of seismic activity between Easter Island and Southern Chile. The American nuclear submarines *Nautilus* and *Skate* reported a rugged mountainous structure along the epicentre line on their way through the Arctic basin to the North Polar ice cap, but although the few echo sounding profiles made showed a central valley, not enough of them were taken to be sure that it was a continuous feature of the ridge. Oddly enough the most prominent mountains in the Arctic basin, the Lomonosov Ridge, discovered by Soviet oceanographers in 1948, are not a seismically active feature. The epicentre line runs parallel to the range but 200 miles to the west.

Short sections of the ridge did not conclusively prove the Ewing-Heezen case but they added up to a very favourable stack of evidence and year by year more information has been piling up. The I.G.Y. had produced a further crop of seismic stations and as they started recording in places like India and the Antarctic, better earthquake epicentre plots were made. Gradually the voyages of research ships have yielded new soundings to extend the hundreds of miles of mountains found in the first exciting days, slowly completing the pattern of what is now called the 'mid-ocean ridge' a continuous system approximately 40,000 kilometres in length. The story of this extraordinary discovery will probably be told as long as people have any curiosity left about the earth we call our home, for these mind-haunting mountains, drowned in the oceans, would be, if we could only see them, the incomparable physical spectacle of the planet.

4

The Biggest Feature on Earth

40,000 KILOMETRES OF NEWLY-DISCOVERED SUBMARINE MOUNTAINS WERE greeted by geographers, geologists and geophysicists with a surprise that rapidly turned to embarrassment. Now that they had been revealed, these great chains of peaks, sometimes 12,000 or 15,000 feet high, marching across the ocean floor, could hardly be ignored. Such a colossal and coherent physical feature not only demanded explanation; by competing in size and scale with the continents and ocean basins themselves, it clearly had to be included in any future conjectures about the structure and history of the earth.

For quite a while the theoreticians had very little help in the way of visual aids except the earthquake plot and some profiles drawn from transits across the ridge. Eventually Marie Tharp and Bruce Heezen produced their famous 'physiographic diagrams' of ocean basins incorporating not only the soundings plotted on the GEBCO charts but their own new rifted mountain chains as well, sketched in simulated relief. These diagrams, pre-empting a fair amount of wall-space wherever oceanography is undertaken, are the graphic summary of two centuries of oceanographic effort. They are the basis for virtually all available paintings and sculptured maps of the ocean floor and represent a sizeable scientific investment as each one took between two and five years to complete, even when a computer was brought in to help. Revising them in the light of still later discoveries is a daunting task.

Drawn out in partial elevation in the diagrams, the Mid-Atlantic Ridge, the first part of the system to be discovered, is distinctive because for most of its length it divides the whole Atlantic basin so exactly, its long curves marking a ghostly outline of the distant continents. Swelling gently out of the flat abyssal plains on either side, the ridge at its broadest parts is 1,000 miles across. The deep median valley and its bordering lines of high peaks show up as a sharp and almost continuous line for long distances although it is broken into innumerable offset sections by the fracture zones that split it from side to side. The effect has been well compared to a long French loaf that has been sliced and then moved so that the slices no longer line up together.

The volcanic peaks like the Azores, Tristan da Cunha, St. Helena and Ascension Islands that break the surface are all, with the doubtful exception of the St. Peter and St. Paul Rocks in the central Atlantic, not the peaks of the ridge but eruptions on its flanks and there are a lot of similar volcanoes and conical seamounts that do not rise above the water. In the North Atlantic the main part of the ridge makes a sudden long offset to the west to continue as the Reykjanes Ridge with an unusually long straight section, continuing northwards into the Arctic after the strange and apparently significant interruption caused by Iceland.

At its southern end the whole ridge swings round the Cape of Good Hope into the Indian Ocean where its shape becomes more complex, looking in general outline like a huge inverted letter Y. The tail of the Y curves northwards, suffers a big offset at the Owen fracture zone, then carries on into the Gulf of Aden and through an almost 90° turn into the Red Sea. The complexity of this part of the ridge system is emphasised by the pattern of rift valleys that scar the continental rocks in East Africa and stretch from Mozambique to Jordan. In the Eastern Indian Ocean the charts show a long isolated section of what looks very like the same kind of ridge, the Ninety-East Ridge, running slightly west of south from the Bay of Bengal. This, whatever it is, is not a seismically active stretch of the main ridge system, and is now usually interpreted as a fracture zone.

The main ridge curves out of the Indian Ocean midway between Australia and Antarctica until it enters the South Pacific and, in a series of offsets, curves north-east again into the wide and gentle contours of the East Pacific Rise. The rise throws out occasional short branches sideways but in general it echoes the outline of western South America, the high crests of the Andes dropping down in an almost straight nine-mile plunge to the deep ocean trenches that hug the coast. Finally the great mountain chain of the East Pacific Rise seems to bury itself under western North America, leaving a zig-zag trail of small ridge sections in the Gulf of California and, in the most recent maps, another scattering of short sections, looking distinctly out of position, to the north of Cape Mendocino. From the crest of the vanished ridge, continuing the broken pattern of the

EAST TO WEST PROFILES ACROSS THREE
OCEANIC RIDGES.

Arrows show position of median rift.

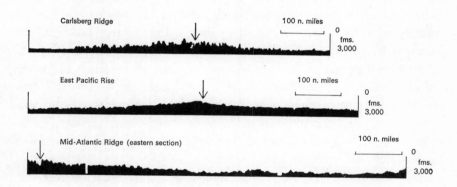

whole structure, the eye picks out the long straight lines of the fracture zones, stretching out westwards into the confusion of volcanic island chains, guyots, atolls, submerged seamounts and great trenches that make up the topography of the western Pacific.

Before long the pattern of the mid-ocean ridge system will be part of the mental furniture of every child who goes to school and even before its full world-wide extent had been charted there were theories about its significance, especially about the first section to be identified, the Mid-Atlantic Ridge. It was not long before some romantic traditionalists were hailing it as the remains of Atlantis, the lost continent described in convincing detail by Plato in two works, *Timaeus* and *Critias*, written sometime around 360 BC. The story has always been regarded in modern times as a legend, though Plato treated it as history, handed on from the records of the Egyptian priests to Solon, the great Athenian lawgiver and passed on by him as a family tradition to Plato's own contemporary, Critias. According to this ancient story the rich, powerful civilisation of Atlantis, the forerunner of historic Greece, vanished into the sea in a day and a night after cataclysmic floods and earthquakes, carrying with it the whole high culture of temples, cities and valiant fighting men.

Now here at last was the Atlantic giving up its secret, the huge backbone of mountainous land that had foundered some ten thousand years ago. The suggestion had all the emotional appeal of fiction proved to be fact, the shadowy traditions of pre-history vindicated by modern science; but although it was a nice idea it was by no means the only one. Naturally enough the Mid-Atlantic Ridge was soon featuring in the heated debate that raged around the subject of continental drift. According to one school of 'Drifters' the ridge was a fragment of the original supercontinent, left behind when the present-day continents drifted apart. Not so, said another school: the ridge was built out of uplifted sediments that had filled the great crack left in the crust when the continental fragments broke loose.

Neither hypothesis was founded on a shred of evidence about the nature of the ridge itself, whatever the value of the continental drift theory, and both of them were demolished by the evidence from seismology and dredging. If the ridge had been a continental fragment, earthquake or seismic waves ought to have travelled through it at the low speed characteristic of light continental rocks.

If it was made of sediments the speed should have been lower still. Instead it was 6–6·7 kilometres per second, the rate characteristic of oceanic crust and only in the median valley was a slower rate recorded. Similarly the rock samples dredged up ought to have been either the volcanic, granitic rocks typical of continents or sedimentary rocks laid down in water. Instead they were mainly basaltic lavas mixed up with rocks like peridotite, serpentine and gabbro, the sort of rocks that the earth's mantle is expected to contain.

From the beginning it was plain that the mountains of the mid-ocean ridges were quite different from any of the known types of continental mountains. Even if many rock samples dredged up from the ridges were basaltic lavas, very similar (except for the quick cooling effects of cold water) to those of land-based volcanoes, the echo profiles showed that this ridge was unlike any other type of volcanic range. In continental ranges like the Cascades and the Andes, rows of individual peaks stand partly buried in their own debris, but these submarine peaks presented an altogether more organised appearance, descending in symmetrical and ordered ranks from each side of the great central rift. Nor were they like the ranges of 'folded' mountains such as the Appalachians in the Eastern United States or the Jura in Northern France which suggest wrinkles in stratified rocks originally laid down as sediments 10 miles thick. In comparison the central valley and the surrounding volcanoes of the mid-ocean ridge are remarkably free of sediments although a few metres of coarse material seem to have settled occasionally in 'ponds' and pockets. Nor did the new ridge mountains, their individual peaks often elongated parallel to the axis of the chain, seem to have much in common with the great shattered blocks of the faulted ranges like the Rockies. Nor indeed did any of these familiar continental mountains have a central valley quite like that of the Mid-Atlantic or Indian Ocean Ridges.

For a century everyone had more or less accepted the explanation that both folded and faulted mountains were the results of compression of the earth's crust, probably due to shrinkage of the surface skin as the hot planet cooled down; but these mountains, with their central valley, were apparently related to a totally different kind of structure, to the great unexplained tension cracks of the rift valleys in the Levant and East Africa and the depression forming the deep Central Icelandic 'Graben', cracks that implied that the surface rocks had been pulled apart until they split, by gigantic forces working at right-angles to the axis of wide valleys. And from analysis of the records of volcanic activity there was evidence that the rifts were gradually widening. In Iceland, where measurements could be taken fairly easily, new volcanic rocks forming along the cracks seem to be spreading the central rift apart at a rate of 3·5 centimetres per 1,000 years for every kilometre of its width.

Another set of observations had been made by taking oceanic heat flow measurements. The pioneers in this technique had been Sir Edward Bullard of Cambridge, A. E. Maxwell and Roger Revelle, and by 1956 measurements taken on various stretches of the ridge crests had revealed abnormally high heat flow at places suggesting vulcanism of some kind at work.

Finally, and perhaps most significantly of all, the rocks dredged up from ridges were all relatively young, around 10 or 12 million years old, young enough to suggest that the ridges themselves were new features.

Indeed nothing had been brought up out of the sea – fossil or sediment or rock – older than at most 180 million years, whereas rocks had been found on land that were dated at 3,000 million years old or more. The suspicion (voiced occasionally by people like chemists who said the sea was not salty enough) that the ocean basins themselves might be young, began to harden into belief.

With all this to support him Bruce Heezen of Lamont, one of the original prophets and discoverers of the whole ridge system, proposed, as early as 1960 in writing and maybe as much as two years earlier in conversation, a bold explanation of the ridge that was at least highly imaginative in its approach. Along the line of the mid-oceanic rift that wandered along the crest of so much of the ridge, the surface of the earth was, he said, cracking open, 'coming apart at the seams' and growing steadily larger in circumference as new material from the mantle rose up to fill the gap. He also thought that perhaps the continents were the broken fragments of a complete shell of granitic rock, formed out of light materials that had floated to the surface as the younger, smaller earth heated up into a molten state, and had then solidified to cover the whole surface. At a later stage this shell was broken apart by the mid-ocean ridges which, as they grew, widened the cracks into broad ocean basins and the centre of the ridge would therefore be the thinnest and youngest part of the earth's crust. To support this idea Heezen painted the continents and the mid-ocean ridges onto a globe and showed that without the ocean basins they more or less fitted together at their margins into a continuous shell of land.

At the very mention of this theory a number of people told Heezen that he was crazy. A shrinking earth was tolerable but an expanding one was totally unacceptable.What you can believe depends, it seems, on what you are used to and although future geophysicists will probably look back on both proposals as equally dotty the 'shrinking earth' theory had been comfortably established for over 200 years. It was first suggested by Sir Isaac Newton and later, during the nineteenth century, it was calculated that an initially hot molten earth would have taken 100 million years to cool to its present temperature, at the same time contracting its circumference by tens or even hundreds of miles.Why the cooling and contracting should give rise to an irregular pattern of continents unevenly distributed across the globe was not explained. It was assumed that the granitic blocks of continental rock had somehow differentiated out from the molten material and had been frozen into their present positions.

The expanding earth notion wasn't just a flight of Heezen's fancy; it also had some kind of ancestry behind it in a paper read in 1956 by Professor S. W. Carey of the University of Tasmania, at a symposium on continental drift, when the whole mid-ocean ridge was still in the process of being discovered.

Twenty-five years earlier the British physicist P. A. M. Dirac had proposed an expanding earth on the cosmological grounds that the force of gravity decreased in proportion to the age of the universe. More recently R. H. Dicke of Princeton had calculated that the gravitational constant would have decreased enough to allow the circumference of the earth to increase by 1,100 miles over 3·25 billion (325,000 million) years, a sum which produced some dignified support for the Heezen theory.

Heezen was not howled down by his colleagues; there was too much revolutionary thinking in the air for anything so uncouth to be wise, or even necessary. Oceanography simply moved onwards so fast that the expanding earth theory seemed to vanish as fast as it had appeared, trampled to death in the rush of new ideas.

In its place came the skeletal outline of the theory that now holds the ground; the theory of sea floor spreading, which has been added to, revised and reinterpreted often enough in the last few years to make it look like an acceptable first draft of the true history of the earth's crust. The sea floor spreading idea was published in 1962 more or less simultaneously from two sources, by Robert Dietz, of the U.S. Naval Electronics Laboratory, who gave the theory the name that has stuck and later in greater detail by Professor Harry H. Hess of Princeton though Dr. Dietz has always conceded that Hess had arrived at his theory, was discussing its implications and even lecturing on it a year or two before he got around to publishing his paper.

Hess decided that the earth was *not* expanding by the addition of new material at the mid-ocean ridges. He considered that to be a 'philosophically unsatisfying' idea. Instead he saw the mid-ocean ridges as one end of a huge system which engaged the sea floor in a slow, steady process of overturning and renewal. At the ridges, marked by the line of shallow earthquakes, new molten material from the earth's mantle welled up to fill the tension cracks of the median rift and solidified into rock at the rate of a few centimetres a year, gradually widening the ocean basins. Meanwhile, at the other end of the system – and this was the crucial innovation – the sea floor was being equally constantly *destroyed* by descending into the deep oceanic trenches and back into the mantle, to the accompaniment of all the deep earthquakes and volcanic activity associated with the trench sites. In such a system the circumference of the earth would not need to expand or to shrink but could remain the same.

Hess based his idea of the constantly moving sea floor, growing outwards from the mid-ocean ridges, on a hypothetical history of the earth. He adopted the contemporary proposals that the planet was formed of some kind of large collection of dust or solid particles which gradually condensed into a smaller compact sphere. Gradually the short-lived radioactive elements in the interior of the growing earth heated it up and

POSSIBLE GEOMETRY OF A MANTLE
CONVECTION CELL PROPOSED BY
H. H. HESS.

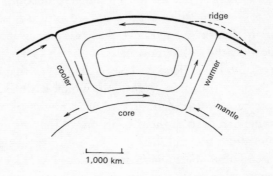

1,000 km.

partially melted it and within the molten interior the heat drove the hot material outwards and upwards in a great overturning movement, allowing it to sink back to the interior when it had cooled at the surface.

This convection system in the form of a single cell might, he thought, have been the cause of a 'great catastrophe'. Its activity might have separated out the light materials such as silicates with low melting temperatures, leaving the heavy materials like nickel and iron to form a core at the centre of the earth. (This idea has recently been extended to include sulphur in the mix of the core.) Eventually the light granitic silicates, the continental-type materials, would be carried over the surface of the globe until they collected above the down-going limb of the cell's current like scum on a hot brew, making a single primeval continent too light to be carried back into the earth. Hess thought as much as 50 per cent of the present continents might have been made at this time, soon after the formation of the solid earth, the remaining 50 per cent being accounted for by the outpourings of subsequent volcanoes.

Instead of being covered by a primary, continuous shell of granitic continental rock as the expanding-earth idea demanded, Hess's primitive earth, from the moment his 'catastrophe' took effect, was divided into two unequal parts, the smaller one containing most of the land and the larger most of the sea, a division known as the 'bilateral asymmetry' of the planet. The ocean water, at about one third of its present volume, would have been produced by the process of rock formation, as the materials of the mantle cooled down and released steam.

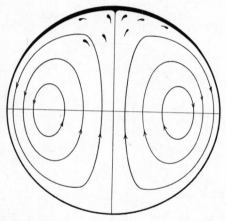

Single cell convective overturn of the interior of the earth before the formation of the core, as proposed by Vening Meinesz and Hess. They thought continental material might have been extruded above the rising limb of the current to collect eventually above the descending limb.

Once all this had happened the existence of the nickel-iron core in the middle of the earth put a stop to the simple single-cell convection system and forced the molten material to form a multiple-cell system instead. The present mid-ocean ridges should therefore mark the rising limbs of several convection cells in which the hot mantle material was rising to the surface at the rate of about one centimetre a year. As it rose towards the surface this hot material would cool down to about 500°C. Its own water would change its own olivine peridotite to a more hydrated rock called serpentine. Also, because the level at which this critical temperature was reached remained constant the level of serpentine formation also remained constant, so that the newly formed oceanic crust became a layer of extraordinarily uniform thickness, approximately five kilometres, established over thousands of kilometres of horizontal distance by seismic studies. This uniformity could not possibly be achieved if the ocean floor was formed by the random outpouring of lavas at the ridge crests.

The convection cells in the hot and viscous material of the mantle rafted the ocean floor away from the ridges like pairs of conveyor belts operating back to back. From time to time, as it left the ridge, the layer of sea floor might be disrupted by volcanoes, some of which might manage to break the surface of the water to become oceanic islands, the oldest ones always lying farthest away from the ridge, while others, less active, would remain unseen below the surface as seamounts.

SUGGESTED MULTI-CELL SYSTEM OF
CONVECTIVE OVERTURN AFTER THE
FORMATION OF THE EARTH'S CORE.

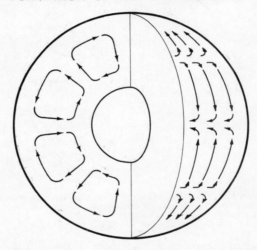

The volcanic cones above the surface would, in the course of time, suffer erosion by weather and waves until their tops were ground down to sea level. Then, as they moved inexorably away from the ridge axis the whole range of peaks would slowly sink, carrying the flattened cones beneath the surface to become guyots, or in the right conditions of sinking rate and water temperature, atolls.

As the sea floor moved along it gradually gathered a covering of sediments, a slow deposit of the shells of minute sea creatures together with other marine debris mixed with atmospheric dusts and volcanic ashes blown off the continents and meteoric particles from outer space, accumulating at a rate usually less than a centimetre in a thousand years.

Eventually the outgoing rafts of sea floor would be carried over the down-going limbs of the mantle convection cells where their leading edges would be drawn downwards and destroyed by remelting to rejoin the hot mantle. There, in a splendid phrase, 'The cover of oceanic sediments and seamounts ride down into the jaw-crusher of the descending limb' where, thought Hess, they might be metamorphosed and welded onto the edges of the continents. In this way the whole ocean would be replaced every 300 or 400 million years, swept clean so that no traces of older rocks or sediments would be left to be found and examined by oceanographers.

FORMATION OF GUYOTS AND ATOLLS PROPOSED BY H. H. HESS.
Volcanoes forming on the ridge crest migrate laterally with the crust and gradually submerge. Volcanoes A and B in Profile A become paired guyots, their tops planed by wave action, in Profile B. Volcano C is younger than volcano B and has formed on its site in Profile B. As it moves away and sinks it also becomes a guyot but has a higher summit than B.

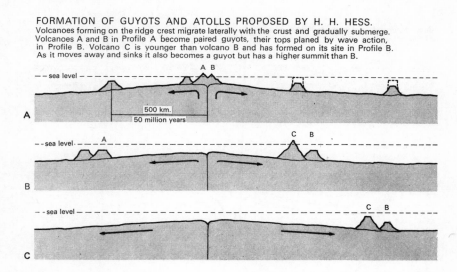

If the rising limb of a new convection cell came to the surface under a continent it would break the continent apart, separating it centimetre by centimetre until after millions of years the spreading ridge of an ocean like the Atlantic would be found perfectly placed between symmetrical sections of sea floor. The broken continental fragments would be carried passively along on the convecting mantle until they reached a down-going limb of current and there they would stop. Because of their lower density relative to the heavy basaltic sea floor rocks they would be unable to move downwards themselves into the mantle, though their leading edges would be thickened and deformed, or even buckled into the great mountain chains like those that run along the western coasts of North and South America, to the accompaniment of earthquakes and volcanoes.

Hess offered his famous 1962 paper as an 'essay in geopoetry', trying not to travel too far into the realms of fantasy. In fact a great deal of it was a brilliant creative rearrangement of many different ideas acknowledged to have come from other people. One of his debts was certainly to the great British geologist Arthur Holmes, of the University of Edinburgh, who in 1931 had proposed a mechanism of convection currents in the mantle. To explain the formation of the oceans by thinning of the earth's crust, Holmes was the first person to see the deep trenches as the places where the currents in the mantle descend again into the interior of the earth. Farther back on the chain of ideas, Holmes himself had precursors in the con-

Continental drift and new 'stretched' ocean floor produced by the agency of convection currents in the mantle, were visualised by Arthur Holmes in 1931

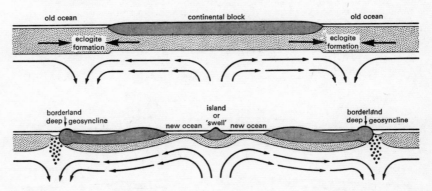

vection idea; one of them, Hopkins, had published a paper suggesting crustal convection as early as 1839!

Another of Hess's contributors was the Dutch geophysicist, Felix A. Vening Meinesz, who had specialised in the measurement of gravity anomalies over the ocean floor. He got his results by managing to make pendulum gravimeters work in submarines and he found that the deep ocean trenches showed such strong 'negative gravity anomalies', in other words, gravity deficiencies, that he proposed thermal convection in the earth's mantle as an explanation for a force that seemed to be pulling down the sea floor into these depressions and preventing it from following its 'natural tendency' under the force of gravity to flatten out.

As for Hess's ideas on continental drift, it is no longer possible to say that they came from du Toit, Wegener and Taylor, its most famous apologists. Now that the subject has become respectable, it trails behind it a whole genealogy of people claimed to be its originators, even stretching back as far as 1620 when Francis Bacon, in his *Novum Organum*, writing on 'Regularities in Nature', said:

> ... The very configuration of the world itself in its greater parts presents Conformable Instances which are not to be neglected. For example Africa and the region of Peru with the continent stretching to the Straits of Magellan, in each of which tracts there are similar isthmuses and similar promontories which can hardly be by accident.
>
> ... Again, there is the Old and New World, both of which are broad and extended towards the north, narrow and pointed towards the south!

Bacon certainly did not say in so many words that the continents might have been joined together in the past though he seems to be hovering just one thought away from the idea. Perhaps a better claimant is François Placet, a French moralist who belonged to a curious order called the Premonstratensions, and published a booklet in 1666 about sin and its consequences under the title *La Corruption du grand et petit Monde*. Starting from the conventional belief that the flood of Noah was a divine punishment for human wickedness, Placet took off on an idea of his own and indulged in a continuation of the title which translates 'Before the Deluge America was not separate from the other parts of the Earth, and there were no Islands.' Although this sentence guarantees an obscure religious eccentric unexpected immortality in the history of science, Placet should not be taken too seriously. His big idea amounts to little more than a speculation that America was formed either by a collection of floating islands or by the uplift of new land to compensate for the sinking of the legendary Atlantis.

Theodor Christoph Lilienthal, an eighteenth-century Professor of Theology in the University of Königsburg, positively supported the literal statement that the earth was divided by the Flood and cited the 'congruent shape' of South America and Africa to prove it. In 1801 Friedrich Heinrich Alexander von Humboldt explained the fit of the coastlines and the geological similarities of the continents by saying that the Atlantic Ocean is 'nothing more than a valley scooped out by the sea'. Each claimant emerges from the shadowy past of scientific literature, at the bidding of his erudite, industrious sponsor with something very like a delicate whiff of incense, the faint echo of a call for canonisation.

Much later, Antonio Snider-Pellegrini, the first of the moderns, with the young sciences of geology and palaeontology to help him, wrote in 1858 in French a strange book the title of which can be translated *The Creation and its Mysteries Revealed*. He suggested that all the continents must have been joined together in order to explain the similarities in the American and European rocks and fossils, especially plants of the Carboniferous period and published a diagram to make the point. He also thought that the American continent had been moved to its present position by an earth-shattering catastrophe, remembered as the flood of Noah. The catastrophe itself was due to instability of the globe when all the continents were clustered in one side of it. The instantaneous translation of America to the other side of the world re-established equilibrium. It is all fascinating stuff, not quite fanciful enough to qualify as fiction, and less astonishing than fact.

Undoubtedly a great proportion of Hess's insight sprang from the results of his own work in the western Pacific in World War II when his deep echo sounder located the mysterious landscape of flat-topped guyots hidden below the atolls and islands that dot the surface of the sea. His papers on this work show how much thought these strange structures provoked. In the course of reaching his conclusions on their 'life-cycle' from volcanic islands to wave-eroded platforms to sunken relics his mind must have been forced into reconsidering the whole sweep of marine geological theory.

Strong corroborative material was supplied by the results of palaeo-magnetic studies of continental rocks by people like Professor S. K. Runcorn, of the University of Newcastle, and his collaborator Dr. E. Irving, whose findings could only be explained if the continents had moved over large distances in geologically recent time. Meanwhile Professor M. W. Menard at Scripps was pouring out ideas about such matters as the rate of rock production by volcanoes and the fact that while the Mid-Atlantic Ridge and the ridges in most other oceans lay along the median lines of the basins, that of the Pacific, the East Pacific Rise, was, as its name implies, markedly east of the middle of the ocean. Menard had explained this by proposing that while the other ridges had formed under continents and so had their symmetry 'built in' as they spread, the East Pacific Rise had opened up in an existing ocean basin and had therefore never been in a central position between continental margins. Menard and Hess agreed that ridges might be ephemeral features rising and spreading, then dying and sinking away again over a relatively short geological time-scale, and that a hypothetical 'Darwin Rise', active in the north-west Pacific until 100 million years ago, would in subsiding into inactivity account for the dense scattering of Pacific islands, atolls and guyots in that region.

POSITION PROPOSED FOR THE
HYPOTHETICAL 'DARWIN RISE'.

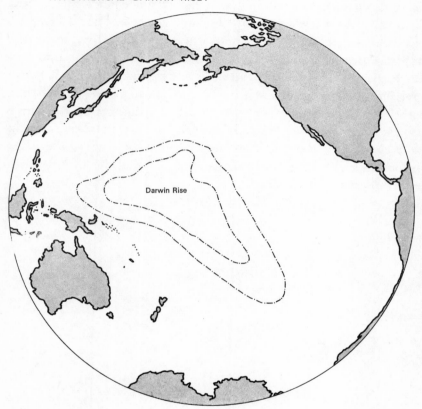

Darwin Rise

Finally it would be unjust to omit the influence in Hess's great geo-physical construction of Heezen's first flush of interpretative ideas. He it was, after all, who first formalised the notion of the rifted ridges, the con-stant growth of the sea floor and thereby supplied a possible mechanism to account for the separation of the continents without obliging them to 'sail the ships of sima through a frozen sea of sial' presumably leaving in their wake a great and so far undetectable scar in the deep ocean basins. Un-luckily for him, he appears to have missed seeing the significance of the consuming trenches. Thinking only in terms of the spreading ridges he knew to be at work in all the oceans, he was obliged by logic and a sea floor surplus to reject the idea of continental drift on an earth of constant size, and in science, alas, as elsewhere, there are no prizes for being runner-up – the winner takes all.

The sea floor spreading idea offered by Hess (and, in a slightly different formalisation, by Dietz) certainly brought new guidelines for the research efforts of all earth scientists. Not all geophysicists welcomed them, and land geologists, in spite of the fact that it was from them that the demand had first arisen for an acceptable explanation for continental drift, were now particularly unmoved by this challenging new framework. Whatever was going on in the oceans did not at first sight seem to have much bearing on their small-scale problems of rock fractionation and chronology. In any case, except for the ridges themselves, there was nothing that, for a geologist of conservative disposition and settled habits, could be called evidence to support such a wholesale overthrow of the accumulation of more than a century of painstaking work.

5

All Change!

By pausing at this precise point in the story of geophysical discovery in the oceans, it is possible to catch a glimpse of the devious and serpentine advances of science itself.

There is a traditional children's game called Grandmother's Footsteps, in which the field of players advances stealthily toward one opposing player whose back is turned to them. The aim of the field is to step forward as far and as fast as possible without being seen, while the aim of the lone player in front is to turn round quickly enough to catch any of them in mid-movement and force them to return to their starting position. Sooner or later, the front player meets his moment of crisis: not a movement can be seen yet all the players have managed to change position. Trying to catch the steps of scientific advance when they are actually taking place can be a little like playing that game. By the time the front player, the observer, has realised that something interesting is going on, the move has been made, the work has been done, the paper has been published and while no-one was looking the positions of all the workers in the subject have altered.

Looking backwards now to the beginning of the 1960s there was clearly a moment when a whole generation of scientists in oceanographic research were just about to step forward; a step so long and well recorded that we can almost replay the action scene by scene to watch the significant movement of the individuals concerned.

The first scene opens as long ago as 1952, when a British scientist, Ronald G. Mason, working at the Scripps Institution of Oceanography at La Jolla, tied a magnetometer to the stern of a research ship at Samoa and towed it half-way across the Pacific to San Diego, the home port of the Scripps 'fleet', and for the first time measured precisely the earth's magnetic field over a broad stretch of ocean. The results showed much greater local variation in the magnetism than the topography of the sea floor would have suggested, so Mason and some other Scripps researchers decided to continue the experiment and built a special magnetometer of their own for the job.

The researchers decided to measure the strength of the earth's total magnetic field very precisely along a series of closely spaced parallel lines and then by mathematics to separate the signals of the earth's general field from the local or 'anomalous' component due to magnetic rocks and minerals in the earth's crust.

Although quite a lot of this kind of magnetic surveying was going on at the time on land under the strong incentive of the search for oil and minerals, there was no reason except curiosity to apply the techniques to the ocean floor. The curiosity began to wear off as nothing much emerged except a general and unexplained magnetic roughness that could not be correlated with any distinguishable features on the sea floor charts. Just when they were at the very point of giving up the team were offered a chance in 1955 to tow their magnetometer behind the U.S. Coast and Geodetic research ship *Pioneer* which was setting off on a deep-water project off the Californian coast and proposed to steam along east to west track lines five miles apart on a course accurate to within 150 yards.

When they had enough magnetic profiles, Mason, expecting nothing, drew up a 'contour' map, tracing the signals of equal magnetic intensity as lines, in the same way as the lines of equal barometric pressure are drawn on a weather map. The result was startling. The magnetic 'anomalies' ran more or less parallel with each other alternating between positive and negative bands running in a north-south direction right across the map, making a pattern far more regular than anything ever seen in a land survey. For the next few months everyone who could be persuaded to take part in the tedious business of looking after the magnetometer on tow was pressed into action and by the end of 1956 a similar uniform pattern of magnetic 'stripes' had been established over a 1,400-mile stretch of ocean floor from Queen Charlotte Islands off British Columbia right down to Mexico.

No immediate explanation of this remarkable phenomenon could be found although there were a number of fascinating incidental discoveries.

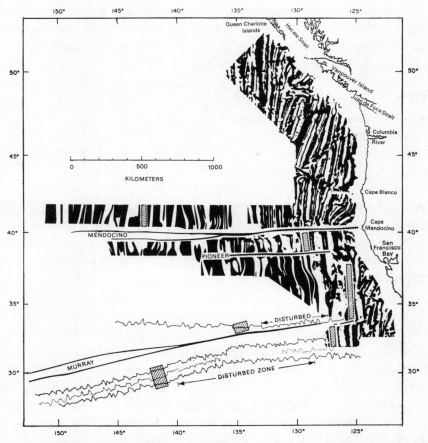

Strange striped pattern in the record of sea floor magnetism off the coast of California. Hatched blocks show matching areas displaced along fracture zones.

From time to time the pattern of parallel lines was sharply broken along an east-west line and several of these breaks turned out to coincide with the lines of great faults discovered a few years earlier in the eastern Pacific floor by sounding, yet no matter how much these magnetic contour lines were moved around they could never be matched up again after the break of the fault. Finally another Scripps worker, Victor Vacquier, suggested that the slippages of crustal material along the faults might be so large that they were beyond the limits of the magnetic map.

So it turned out to be, although it took two years of laborious surveying to prove the point. By extending the survey to the west the match of the magnetic readings was finally established 130 nautical miles along the Pioneer fault, and 600 miles away along the Mendocino fault, revealing a total displacement of crustal blocks of more than 700 miles. Why and how it had happened was a mystery, as were the strange magnetic parallels themselves. The only clue was in the samples of basalt lavas dredged off the ocean floor, since basalt is the most highly magnetic of all the common rocks. As the basalt lava cooled through its Curie point of 500°C, it would retain, frozen into it, the pattern of the magnetic field in which it lay; but why should the pattern change in parallel stripes? Could it be connected with the stresses that had folded the west coast of North America into mountains?

Writing about these discoveries in 1961, Arthur Raff, one of Mason's co-workers, made one last speculative leap and suggested that as the magnetic stripes ran parallel to the lines of the oceanic ridges in the Pacific these two phenomena might be related. Although he did not know it his idea carried him within reach of solving the puzzle, but several vital clues were still missing.

Meanwhile another scene in our action replay was being enacted by three other 'magnetics' researchers – Allan Cox and Richard B. Doell of the U.S. Geological Survey at Menlo Park and G. Brent Dalrymple of Berkeley. From the study of magnetism in rocks it had been known for a long time that some rocks are reversely magnetised with respect to the earth's magnetic field. As early as 1906 it had been suggested that this might mean that in the distant past the earth's field had reversed from time to time, in other words that the present magnetic North Pole had become the South Pole and had then changed back again. The issue was confused for a while by the discovery that some rocks were actually 'self-reversing' in other words they assumed, as they cooled, magnetism of opposite polarity to the field of the earth. It turned out, however, that such rocks are extremely rare.

Plenty of examples of reversely magnetised rocks had been found all over the world and their apparently different ages suggested that the reversal process had taken place not once but several times. Cox, Doell and

Dalrymple set out to find out exactly when and how frequently these reversals had happened, by carefully measuring the magnetism in rock
samples and dating them by the potassium-argon method. This technique
of determining the amount of the inert gas argon 40 formed in the rocks
by the decay of radioactive potassium 40 with reference to the known
steady rate of potassium decay is now well established but in the early
1960s although not new it was still relatively untried. To apply it Cox,
Doell and Dalrymple chose a research site where successive layers of
volcanic lavas had erupted from continuously active vents over a long
period and where some of the layers were known to be reversely magnetised. The first results were promising and were published early in 1963,
suggesting that recent reversals had taken place as frequently as every
million or half-million years (due to a quirk of instrumentation it wasn't
quite clear which), and that between each switch there was a period of
indeterminate magnetism lasting around 10,000 years.

It was just at this time, in late 1962, that one of the key figures in the
story, Fred Vine, a young graduate in geology, later a Princeton Professor,
was joining the Department of Geodesy and Geophysics at Cambridge to
embark on a career of research. When he arrived in the department the
man who was to supervise his first research project, Dr. Drummond
Matthews, was away at sea in H.M.S. *Owen* making a detailed magnetic
survey over a central part of the Carlsberg Ridge as part of the work of the
International Indian Ocean Expedition. So for a few weeks Vine was in an
unusually free position to look around for a research project of his own.
As an undergraduate he had heard Hess lecture in Cambridge on his new
theory of sea floor spreading and realised that there was a very strong case
for thinking that continental drift had occurred but no real evidence – for
that geologists were going to have to go back to the ocean.

Part of the business of oceanography is the design of the instruments to
advance research techniques – finding out something new very often
means finding out a new way of looking for it. But Vine was a geologist
and he didn't feel any urge to spend his research time building sea-going
instruments. His interest lay in the rocks of the ocean floor and he wanted
to get at them as soon as possible, so he thought it might be a good idea to
look at the marine magnetic surveys from all over the world and see if they
inspired any explanation of the strange new lineations.

Vine started researching in the literature and one of the papers he
looked at was from Victor Vacquier at Scripps, describing a computer

technique for determining the magnetism of a seamount. He decided to try this method out by applying it to some of the magnetic measurements brought back by Drummond Matthews from the Indian Ocean. Picking out two likely seamounts from the topographic chart, he borrowed a three-dimensional computer programme worked out by K. Kunaratnam, a mathematician at Imperial College, begged some computer time from the mathematics department at Cambridge, got the programme translated into the 'language' of that computer and finally set to work. After all that he was rewarded by finding that one of his two chosen seamounts was normally magnetised and the other reversed. With the implications of this result a whole set of disconnected ideas fell into an ordered arrangement – the mysterious magnetic stripes, the frequent reversals of the earth's magnetic field, the seismically active median rift in the mid-ocean ridges, the frequent samples of basalt dredged from the bottom, and the hypothetical mechanism of sea floor spreading.

If the reversely magnetised seamount had erupted at a time of reversal of the earth's magnetic field, it was possible that the strange magnetic lineations or 'anomalies' of the ocean floor were also produced by rocks cooling in alternately normal and reversed magnetic fields. Vine and Matthews compared the observed magnetic lineations of sections across the Carlsberg Ridge with a set calculated on the arbitrary assumption that blocks of sea floor about 20 kilometres wide were alternately normally and reversely magnetised. The observed 'anomalies' and the calculated 'anomalies' matched. Furthermore, magnetic surveys across other stretches of ridge in the North Atlantic, the Antarctic and Indian Oceans all showed a similar pattern of stripes, all starting from a single strong 'anomaly' over the centre of the ridges.

On the basis of these findings Vine and Matthews published in September 1963 a now famous paper, 'Magnetic Anomalies over Oceanic Ridges', in which they proposed a method of testing the hypothesis of continuous creation of the sea floor offered by Hess and Dietz. If the sea floor was spreading out from the median rifts of oceanic ridges, blocks of alternately normal and reversely magnetised material, probably basalt lavas, would drift away from the centre of the ridge and parallel to its crest. The farther away from the ridge they were the older the rocks would be. Naturally the central strip composed of a number of newly injected dykes would bear the strongest magnetic signature because as they moved away down the flanks of the ridge the blocks of lava would be disturbed by subsequent volcanoes and fissures erupting through them and by more recent lava flows running over them and confusing the signals with their own magnetism. (This explanation of the fall-off in the 'strength' of the magnetism has recently been challenged. It is now thought that a change in rock chemistry due to oxidation might be the cause.)

With the publication of this paper the crucial unseen steps forward in a whole broad front of oceanography had all been taken; 3,000 miles from Princeton the co-ordinating insights of Vine and Matthews had found the

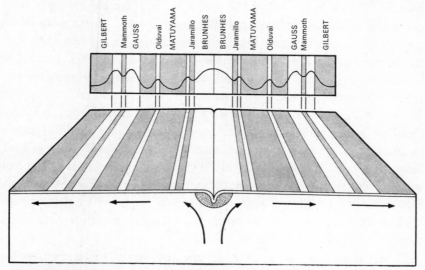

Schematic block of ocean floor with a symmetrical pattern of normal and reversed magnetic lineations as new material spreads laterally from the mid-ocean ridge over the last 4 million years. White strips show normal polarity; grey strips, reversed. Short 'reversal events' appear to break the longer 'epochs' of a single polarity.

way to justify the 'geopoetry' of Hess's ideas and now they could be used in earnest. So it is surprising to discover that the Vine and Matthews paper fell from the presses into a deep well of silence. Although it seems clear today that the geophysical researchers in 1963 were waiting breathlessly for the signal to rush headlong into a new era of exciting discovery, those taking part at the time were aware of nothing of the sort. For more than a year Vine and Matthews had no response at all from either oceanographers, geophysicists or marine geologists outside their own department.

Indeed they were luckier than two other scientists in Canada, Morley and Larochelle, who had tried to publish a paper containing a similar line of reasoning several months earlier. Their ideas had sprung not so much from new field work of their own as from thinking about the results obtained by the Scripps group which had worked on the magnetic survey of the sea floor off California, and where, as we have seen, Vacquier had been finding and investigating anomalously-magnetised seamounts without, however, reaching any conclusions about their significance.

Dr. Morley was the senior geophysicist of the Geophysical Survey of Canada in Ottawa running a huge programme of aero-magnetic surveys but he was also a former student of Professor J. Tuzo Wilson of the University of Toronto, one of the first scientists to respond to the imaginative sweep of Harry Hess's ideas. Morley, then, had exactly the right background to understand the coded message contained in the map of the magnetic lineations (he had found similar patterns in the aerial survey of the Arctic and the North-west Atlantic) and to deduce their general relationship to the field reversals and sea floor spreading. Nevertheless his paper was rejected by learned journals in both Britain and the United States. An anonymous referee wrote to the effect that 'this is the kind of idea that is talked about at cocktail parties but not published in scientific journals'. In great discouragement Morley presented his hypothesis orally before the Royal Society of Canada in Quebec on June 4, 1963. The Vine and Matthews paper appeared three months later. The Morley and Larochelle paper was eventually published in 1964, but by that time the honour of the innovation, so precious to research scientists, was lost to them.

There is another sad story along the same lines. In 1962 an unsung scientist of the U.S. Navy Hydrographic Office, preparing a report on 'Operation Deep Freeze', a general oceanographic survey in Antarctic waters, noting that the magnetic lineations in that area, like those off California, ran parallel to the oceanic rise, wrote that the lineated patterns 'may be an indication of the processes by which the rises were formed'. It is said, rather guardedly, as such circumstances demand, that he wrote quite a lot more on the subject, taking the U.S. Navy even closer to the prize of a major discovery, but the hand of higher authority deleted his 'speculations' from the final report.

Although Vine and Matthews had to wait a long time for acknowledgment things did at long last start to move. Cox, Doell and Dalrymple managed to produce a positive dating sequence for the reversals of the earth's magnetic field for the last three and a half million years. In 1964 Professor Hess spent some time in Cambridge fully aware that the new magnetic work was of the utmost importance. Professor Tuzo Wilson was

AGE OF ATLANTIC ISLANDS

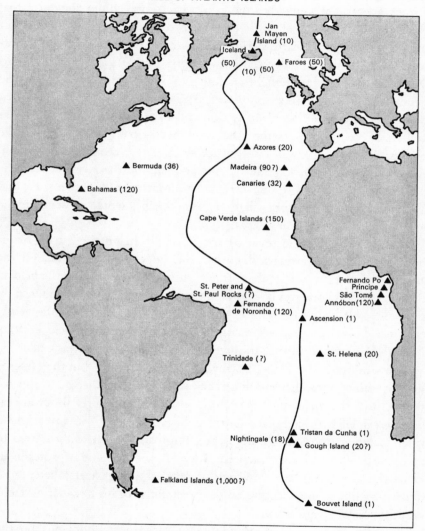

The age of oceanic islands increases the farther they lie from the Mid-Atlantic Ridge. (Figures in brackets represent age in millions of years.)

spending a sabbatical six months in Cambridge at the same time. He too had recently been rebuffed by the learned journals in attempts to publish a paper in support of Hess's theories. Early in 1963 he had written a paper analysing the ages of oceanic islands. He had found that they got older the farther away they were from the mid-ocean ridges and he suggested that the Hawaiian islands could have been formed by the spreading of sea floor over a fracture in the earth's crust from which molten magma was intermittently erupting. Under the encouraging eye of Hess himself, Tuzo Wilson and Vine got together for another look at the seismic and magnetic details of the Pacific floor south-west of Vancouver Island, the northern end of the stretch surveyed by Mason and Raff. Wilson, and Menard as well, had recently come to the conclusion that the complex faultings of the San Andreas system was broken up, once it left the land, by several stretches of spreading ridge. If so, Vine pointed out, they were exactly the kind of thing required to test the Vine and Matthews hypothesis. He and Wilson decided to see if the ridge and its anomalies would substantiate the idea.

They found that the reversed and normally magnetised strips of sea floor spread our symmetrically from the ridge crest and that, on the assumption of a spreading rate of three centimetres a year from the median rift, about 120 kilometres of crustal material could have been produced by the new ridge over four million years, the period covered by the newly established scale of reversals dated by the potassium-argon clock. Vine plotted the lineations against the dated reversals and the case was made.

Now not only the Vine-Matthews 'model for testing' but the Hess sea floor spreading hypothesis took on the respectability of theory. In spite of the fact that the magnetic-reversal time-scale was still fairly shaky at that point it can fairly be said that after the publication of the Vancouver Island ridge results in 1965 the sea floor spreading bandwagon really started to roll. Vine, who had been carried off by Hess to become a professor at Princeton as soon as he completed his Ph.D thesis, gave it a hefty push himself by delivering a lecture on the possibilities of his new dating techniques to scientists at Lamont.

In their oceanographic stronghold in the green woodlands on the Palisades high above the Hudson river, Lamont oceanographers had already 'mapped' beautifully symmetrical magnetic lineations on the Reykjanes Ridge, the section of the Mid-Atlantic Ridge south of Iceland, without attempting to relate them to the Vine-Matthews hypothesis. It

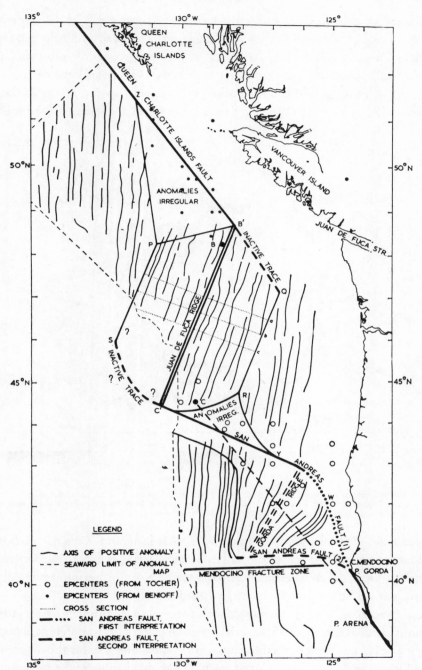

Sketch map by J. Tuzo Wilson of the Juan de Fuca Ridge and its relation with other ridges and transform faults off the coast of North-West California. In this interpretation the San Andreas and other faults are seen as transform faulting between ridge crests.

looked as though they had missed that key paper and the Vancouver Island results as well. If so they soon made up for it. A story is told that the light dawned when the two halves of a flimsy computer print-out of magnetic lineations were accidentally put over a viewing light on top of each other instead of side by side and it immediately became clear that they were almost identical. Whatever really happened, following Vine's lead,

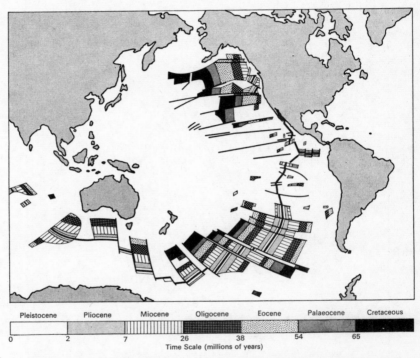

Pleistocene	Pliocene	Miocene	Oligocene	Eocene	Palaeocene	Cretaceous

0   2   7   26   38   54   65

Time Scale (millions of years)

Magnetic anomalies studied in the Pacific make a symmetrical pattern on each side of the spreading edge. The age of the rocks of the ocean floor, increasing outwards from the ridge axis (see key), was established by correlation with identified reversals of the earth's magnetic field.

James R. Heirtzler, W. C. Pitman, G. O. Dickson and Xavier le Pichon found that sections across the ridges in the Pacific, the Atlantic and the Indian Ocean showed exactly the same pattern of normal and reversedly magnetised 'strips' as the Vancouver Island section.

As soon as they had managed, with the help of a computer, to identify individual 'events' of reversal in the lineation pattern of the different oceans they could use them to establish an average spreading rate for the different ridges. There is still quite a lot of argument about changes in the spreading rate from time to time and even a theory that spreading might sometimes stop altogether, and then start again but it seems that in the fastest moving areas, mostly in latitudes south of the Equator, the Mid-Atlantic Ridge is spreading at about four centimetres a year, the Carlsberg Ridge in the Indian Ocean at five centimetres a year and the East Pacific Rise at about 16 centimetres a year, though it may in some places have reached a phenomenal 26 centimetres a year in the past. The high rate of spreading of the East Pacific Rise may account for its flat profile compared with the steeper outline of the Mid-Atlantic Ridge.

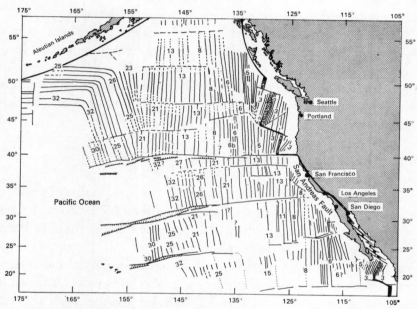

Sketch map of the pattern of magnetic lineations in the North Pacific. Numbers identify anomalies as the signature of reversals of the earth's magnetic field. New plate sections along the coast of California show a change in spreading direction almost certainly related to the overrunning of the earlier ridge crest by the westward drift of North America. The abrupt change of direction south-west of the Aleutian Islands arc and trench system is The Great Magnetic Bight thought to be the remains of the ridge boundary between these earlier plates.

Once these spreading rates are known some kind of time-scale based on them can be extrapolated backwards to date more ancient events. A new jargon has grown up among oceanographers in which numbered 'anomalies' are used to describe both a date and a position in the ocean basin. Anomaly 5 for example was on the middle of the ridge 10 million years ago but it is now a strip of 10 million-year-old sea floor found some distance from the ridge on either side of it, its sections making thin characteristic lines on the map. Anomaly 31 is a similar pair of strips 60 million years old, and anyone who spends a lot of time working with the lineation pattern will get to know where Anomaly 31 lies in different parts of his particular stretch of ocean.

Several large sections of ocean floor have been studied from the ridge outwards in very great detail, especially a long thin area lying across the East Pacific Rise in the region of the Eltanin Fracture zone. In fact the Lamont scientists have more or less committed themselves to the enormous task of dating the whole breadth of the ocean basins by identifying the magnetic signatures of the separate events. After several years of effort they are now fairly confident that they are in sight of the finish.

Unfortunately the reversals of the earth's field have not always been so frequent and regular as they were over the last 70 or so million years. Further back than that there are long 'quiet periods' when there were no reversals and then only extrapolation from the average spreading rates can give an idea of the date of a particular stretch of the same floor. The only way that this can be checked is by positive dating of the sea floor rock underlying anything between 100 and 200 million years' worth of sediment. This is now possible by drilling down and bringing up a rock sample, the kind of work done by the *Glomar Challenger*.

This ship was specially developed for the Deep Sea Drilling Project by the Scripps Institution with a contract from the U.S. National Science Foundation, to solve the problems of drilling into the ocean floor to depths of 1,000 metres, in water up to 6,000 metres deep. She has already completed one 18-month programme of successful operation and is now working on an extension programme for another 30 months. Drilling into igneous rock like basalt wears out a drill bit in less than a metre, so the rock sampling has not been an unqualified success. A re-entry capability has now been developed and the next set of dating results should be very interesting.

Where the drilling has been successful is in recovering sequences of sediments far longer than an ordinary piston corer can retrieve. Drilling on either side of the Mid-Atlantic Ridge has shown that the lowest sediments, those immediately above the basement rock, get older and older as the drilling sites get farther away from the middle of the ridge. Examined by palaeontologists the fossils in the cores have produced dating that fits almost perfectly the dates predicted by the magnetic anomaly pattern.

Here again the re-entry problem has limited the results because the older sediments farthest away from the ridge axis and most in need of exact dating, are also the most likely to have turned into chert, the flint-like substance made of silica which is even harder on drill bits than basalt. The chemistry of chert formation is not fully understood and as not all sediments made of siliceous fossils turn into it, there may be critical conditions that let it form in some oceans or at some times and not at others. It seems for instance to have formed in two main periods, one 60 million years ago and one 100 million years ago, and it seems to be absent in the South Atlantic though widely distributed in the North Atlantic, the Caribbean and the Pacific. Unfortunately the west edge of the North Atlantic, where the *Glomar Challenger* was hoping to find deep, old sediments that

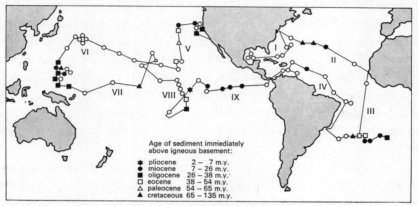

Age of sediment immediately above igneous basement:

★ pliocene   2 – 7 m.y.
● miocene   7 – 26 m.y.
■ oligocene 26 – 38 m.y.
□ eocene   38 – 54 m.y.
△ paleocene 54 – 65 m.y.
▲ cretaceous 65 – 135 m.y.

THE TRACK OF THE *GLOMAR CHALLENGER*

Roman numerals denote legs of her voyages. *Pl*ain circles and symbols show where each hole was drilled in the ocean floor and the age of the sediment found immediately above the igneous basement rock.

would positively date what is thought to have been the first rift in the great continental break-up, has turned out to be particularly favourable to chert, which lies in an impenetrable layer 250 metres above the basement rock. So we still have not got all the answers.

Several years before the *Glomar Challenger* set off on her highly productive drilling voyages, the work at Lamont had perfected another sediment dating technique which can complement the work of the palaeontologists. In 1965, just about the time when the excitement over the magnetic lineation pattern was at its height, a graduate student called John Foster turned up at Lamont and found himself in the department of a palaeomagnetist, Dr. Neil Opdyke. While he was there he succeeded in building an instrument called a spinner magnetometer and with it he managed successfully to identify the magnetic reversal pattern in different layers of sediment. This could be done because minute particles of iron and other things like iron and titanium oxides align themselves with the earth's magnetic field at the time of their deposition. In high latitudes this field makes magnets show a pronounced dip towards the pole. When a specimen of sediment is spun in the machine it behaves like a little bar magnet. The effect is that an alternating magnetic field is being moved round in the middle of a coil and this produces a voltage in the coil and hence a signal that can be amplified and read.

The idea of such an instrument had been suggested as long ago as 1938 to look at the magnetic properties of certain muds that are known to be laid down seasonally in distinct layers. One magnetometer had even been built at the Carnegie Institute but people soon became more interested in measuring the magnetism of rocks than of muds and the rotational speed of the instrument was gradually increased. Although he too was primarily interested in rock samples Foster started again and made his spinner run at slow speed. His real success was due to his managing to reduce the signal-to-noise ratio in the instrument so that the tiny signal from his micro-magnetic particles was not overwhelmed by the different kinds of effective varying magnetic fields, of all kinds of frequencies, that are present in any laboratory.

Late one afternoon, when most of their colleagues were leaving the Observatory, Foster and another graduate student, Billy Glass, went down to the core laboratory and sorted out three sediment samples from a 'high latitude' core bearing the identification mark 'Vema 16/134' from the research ship which had lifted it. They wanted a core of sediments that had been laid down slowly enough for several million years to be represented and that one had already been carefully studied by a palaeontologist so that its various layers were already dated by their fossil contents. This would give them a date correlation for any magnetic anomalies they might find. Everyone had gone home when they started up the noisy magnetometer and spun their three little slices of sediment. Two of them were normally magnetised but the one in the middle was neatly reversed, 'and that,' says Foster, 'was the ball-game'.

What he did not know at that moment was that five examples of reversals had been found in a Pacific core of clayey radiolarian ooze by two other workers, Harrison and Funnell (both British), working at Scripps in 1964. Oddly enough, although the results had been published nobody had shown much interest and the technique had not been developed.

Foster's results, however, produced a frenzy of interest at Lamont. For weeks he was kept busy spinning samples far into the night for research students on different core-dating projects. The result was that a core of sediment could be systematically dated from the top downwards by the dip of the particles as the field reversed, instead of just by the fauna found in each layer. It was possible to use the reversals to compare the ages of

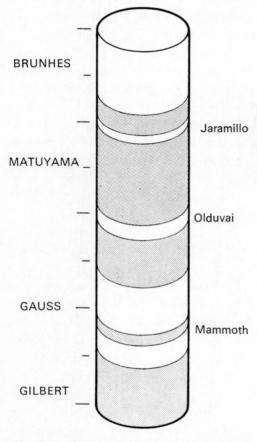

BRUNHES

Jaramillo

MATUYAMA

Olduvai

GAUSS

Mammoth

GILBERT

An oceanic core of sediments records the sequence of polarity reversals by magnetic particles which have orientated themselves in the direction of the earth's field. The names on the left of the core are those of magnetic epochs lasting several million years. Those on the right are shorter 'events' within epochs. The top white layer shows the present 'normal' polarity. The grey layers are reversals.

cores from different positions in the ocean and also theoretically possible to measure rates of sedimentation, though the disturbances caused by machine coring techniques make exact results difficult.

Almost at once it was discovered that the magnetic reversal pattern in different layers in certain cores seemed to coincide with changes in the microscopic fossil fauna. Some species seemed to die out altogether at the time of a reversal and new ones took their place. There was a flurry of speculation, which has by no means died down yet, about what could cause the changes. There were several suggestions. Reversal of the field was thought to be marked by a period of magnetic 'neutrality' of 5,000 or 10,000 years in which unusual numbers of particles and cosmic rays might have got into biologically productive areas of the sea where they might cause genetic changes, instead of being diverted by the earth's field into the high latitudes where they would be less effective. If so, effects of this kind would perhaps turn out to be the explanation of the sudden and mysterious death of the dinosaurs, or the rise of the deciduous plants.

The idea did not really stand up to the argument that the particles would be slowed down by the water in the sea to such an extent that they could not have achieved so much damage in the time available. Another suggestion was that the failure of the earth's field allowed the 'cosmic wind' from the sun to enter the atmosphere and so change the climatic conditions that the balance of success of species was entirely altered. Palaeontologists are now taking a rather colder view of the relationship between reversals and faunal changes. They do not think the changes are quite so sudden and dramatic as the reversals and as they do not *always* coincide there is clearly room for other explanations.

The species changes were not by any means the only cause for speculation. There was also the notion that the reversal of the earth's field was not so much a cause as an effect and that it was triggered by the impact of huge meteorites or perhaps even by the arrival of great showers of extra-terrestrial fragments of glass called tektites which had been found in Lamont cores, appropriately enough by Billy Glass, the graduate student who had helped Foster. This was not a particularly serious suggestion though the coincidence of tektites and field reversals was odd.

Even if tektites are just red herrings in disguise the subject of linked magnetic reversals and biological changes is by no means dead. The trouble is that the cause of reversals in the earth's field is a mystery and so, in an absolute sense, is the cause of the field itself. At the moment, the generally agreed explanation, based on the work of Elsasser and, independently, of Bullard, in the forties and fifties, is that the field is produced by movements in the liquid core acting as a self-exciting, or 'homogenous' dynamo.

The simple analogue of how this works is the disc dynamo, but the picture has now been extended into two disc dynamos coupled together so that the current produced by one feeds the coil of the other. In models this can produce spontaneous reversals of field more or less similar to those of the earth's field, but the validity of this idea partly depends on finding an explanation for the time-scale of the real geophysical reversals. One way and another the subject has acquired such complexity that by people other than fascinated physicists it is called 'arcane' or 'mind-boggling' depending on whether the speaker expects to understand it eventually or has already given up trying.

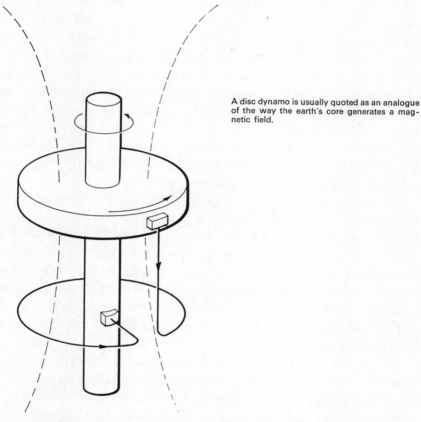

A disc dynamo is usually quoted as an analogue of the way the earth's core generates a magnetic field.

Meanwhile, without rejecting the dynamo idea, another line of reasoning is still trying to link the faunal changes more directly to the reversals without calling on the cosmic wind or elaborate changes in the upper atmosphere. This argument starts from the point that there is some support for the idea that large earthquakes (defined as 7·5 or greater on the Richter scale), account for at least part of the 'wobble' of the earth's spin axis. The wobble itself was discovered by Chandler in 1891 and is properly described as the movement of the earth's axis of figure about its axis of rotation, and if earthquakes can be proved to affect that, then some similar related effect might be enough to cause the magnetic field to flip as well.

New evidence from analysis of cores from the Southern Pacific suggests that at the time of magnetic reversals and coincidental changes of fauna there was a maximum of volcanic activity, spreading out enough dust and ash to be traced in the cores. Dust and ash on this scale would cause major climatic changes by cutting off the solar heating in the higher latitudes where the normal atmospheric circulation would carry it. This sounds like an advance but there is still no positive link between volcanoes, earthquakes and field flip, so the subject is wide open to conjecture!

Although observations of different kinds of phenomena on the surface of the earth may lead to the correct explanation of the reversal problem, there may be a limit to what can be inferred, since it is difficult to imagine how the deep interior of the earth could be investigated. The solution may have to wait for a few more years until computational physics has developed the mathematical techniques to handle the complex equations involved in a model of the earth's behaviour.

During the last four or five years, while the sea floor dating work has been going on, the whole climate of scientific opinion in which it was begun has swung overwhelmingly towards support for continental drift. The first clear sign of a change of mood was the London symposium on the subject sponsored by the Royal Society in 1964. Even the introductory remarks of the chairman, Professor Blackett (now Lord Blackett) cited the build-up of palaeomagnetic and other evidence to suggest that the material question was not 'Have the continents drifted?' but 'How much have they drifted and when?'

Among the papers read on such matters as geological evidence of drift, polar wandering curves, viscosity of the mantle and so on, came a dramatically effective presentation by a Cambridge group consisting of Professor Sir Edward Bullard, J. E. Everett and A. G. Smith. They assembled all the continents bordering the Atlantic into a single super-continental land mass.With the help of a computer to find the best fit, and treating the 500 fathom line, the central depth of the continental slope, as the true continental margin, they managed to close the ocean to within 1° of error. Faced with this fit of the continents, not on a map but rotated across a globe, anyone rejecting the probability of continental drift suddenly found himself called upon to defend his position.

When the full ocean basin dating has finally been achieved there is

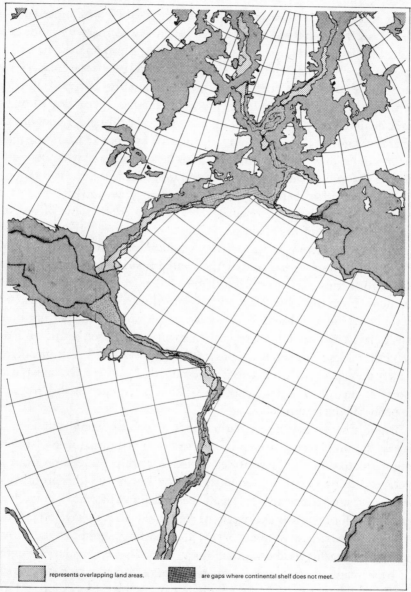

represents overlapping land areas.

are gaps where continental shelf does not meet.

Computer-aided assembly of the continents around the Atlantic at the 500 fathom line, by Bullard, Everitt and Smith in 1965.

bound to be a spate of new interpretative thinking about the whole spreading and drifting pattern. Unfortunately one of the most geologically crucial sections of the ocean floor, from the point of view of understanding what has been going on, is the North Atlantic and there the floor is so broken by faulting that the pattern of lineations is proving the hardest to detect or co-ordinate. The most straightforward situation seems to be in the South Atlantic, the ocean where even the uninformed eye can detect the match of the continents. At the moment there is a great deal of guess-work in all discussions of the timing of almost all drifting episodes but the general opinion is that North America started to separate from Northern Europe and North Africa in the later Triassic period somewhere between 200 and 225 million years ago. The opening of the Labrador Sea is thought to have taken place between 90 and 65 million years ago after which its spreading ridge stopped spreading, allowing the present line of division between Greenland and Europe to open up later on. South America parted company from Africa about 135 million years ago. The next stage was the break-up of the rest of the southern supercontinent of Gond-wanaland. Africa, perhaps still attached to India, separated from the rest of the block. Next India left Africa to travel northwards to Asia perhaps 80 or 90 million years ago leaving Antarctica and Australia together until per-haps as late as 40 million years ago when finally they too separated from each other. New Zealand is variously thought to have left Australia before or after the separation of Antarctica. Because the Southern Ocean has been poorly studied these dates for the separation of Antarctica and Australia are probably the least reliable.

It might seem that the establishment of a positive dating technique for at least the most recent stretch of geological time and the rule-of-thumb dating for the more distant time, by extrapolating backwards from an average rate of spreading, would more or less tie up the problem of whether or not continental drift had taken place and the way it could have happened. But there are, as the critical reader will already have observed, a number of loose ends in the story. While charting the horizontal move-ment of continents has been going on at a great pace very little progress has been made in establishing how the proposed convection cells in the mantle could work. Nor has a satisfactory account been given of the nature of the tension that splits the median rift apart, whether it is pri-marily due to convection or not, nor even of how a trench opens up in the solid sea floor. All these points are likely to feature very largely in oceano-graphical debate over the next few years.

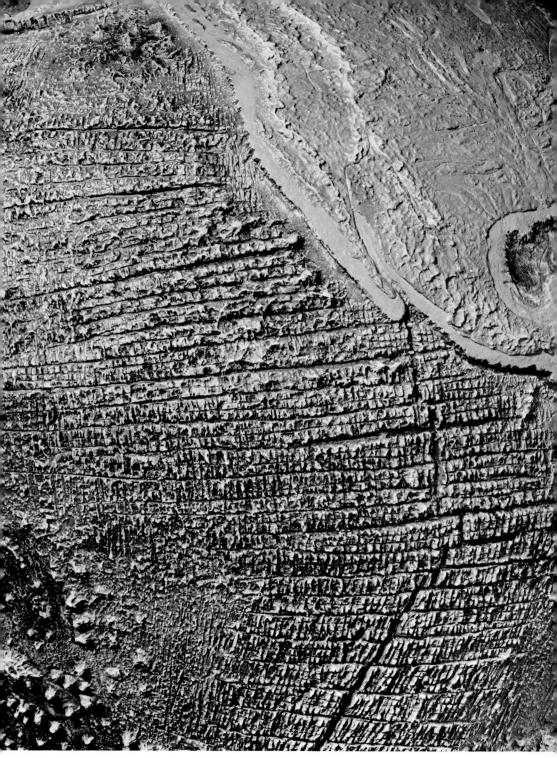

Relief-modelled globe gives an astronaut's eye-view of a waterless ocean. The East Pacific Rise vanishes into the Gulf of California as if overrun by North America. Parallel lines are fracture zones caused by 'transform' faulting between offset sections of the spreading ridge. Fine details of ridge sections in the California area are omitted. (*Photo: Institute of Geological Sciences. Crown copyright reserved.*)

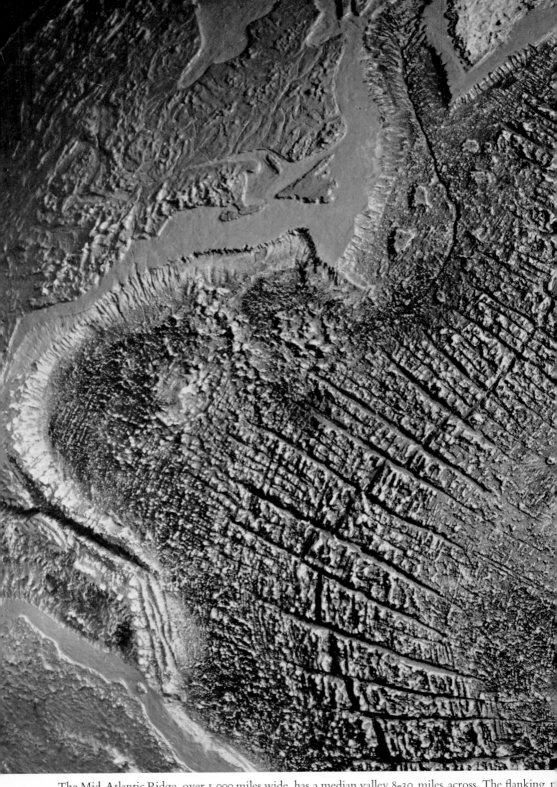

The Mid-Atlantic Ridge, over 1,000 miles wide, has a median valley 8-30 miles across. The flanking p
picture Iceland straddles the ridge line above the straight section known as the Reykjanes Ridge.
is the continental shelf. On the left, in the Caribbean, is the dark cleft of the Puerto Rico Trench

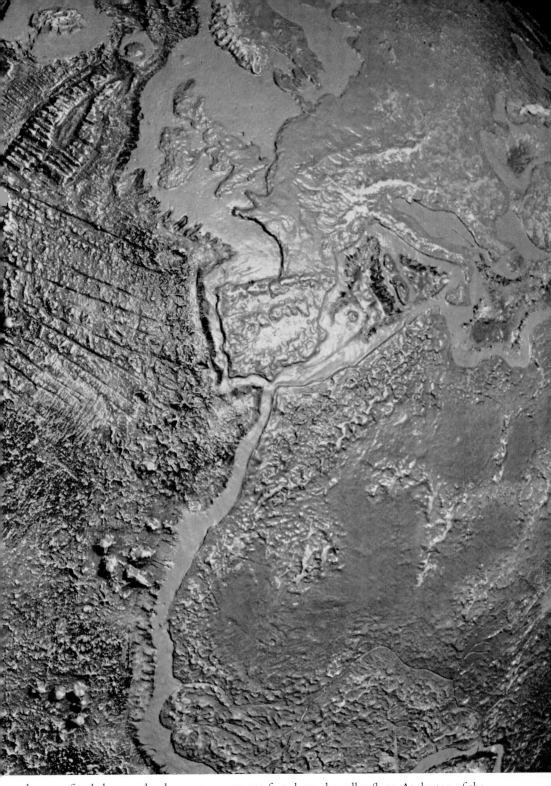

...e only 3,500 feet below sea level, may tower 12,000 feet above the valley floor. At the top of the ...outside the ridge area are the abyssal plains, and the smooth light grey at the continental margins ...test depth of the Atlantic. (*Photo: Institute of Geological Sciences. Crown copyright reserved.*)

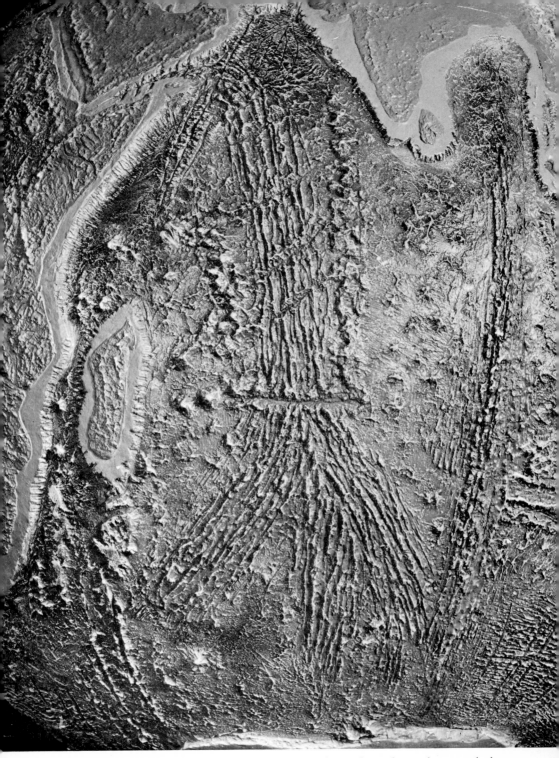

In the Indian Ocean the Carlsberg Ridge runs down from the Red Sea and the Gulf of Aden to split into an inverted Y. The straight range of mountains on the right is the seismically inactive Ninety-East Ridge. (*Photo: Institute of Geological Sciences. Crown copyright reserved.*)

6

Paving Stones and Plates

JUST AS PEOPLE WERE BEGINNING TO GET USED TO THE NEW OCEANOGRAPHIC imagery – the spreading sea floor emerging from mid-ocean ridges as if it was on 'a pair of conveyor belts moving back-to-back' while the successive dykes of basalt preserved the reversals of the earth's magnetic field 'like a slow-moving tape-recorder' – the whole scene was convulsed by a new theoretical advance. This was the birth of the plate theory, now usually called 'plate tectonics'. It describes the earth's crust in terms of a few large moving sections called 'plates' which are bounded by the great faults, the mid-ocean ridges and the deep trenches with their associated coastal mountains and island arcs, in fact by all the most important physical features of the planet. Although attempts to explain the theory are often shrouded in confusion it is really an elegant simplification of a lot of the earlier ideas.

'Plate tectonics' has been so well received and so rapidly assimilated into the mainstream of thinking in marine geology and geophysics that its premises, unheard of until 1967, are now taken for granted in every paper that is published. All the same they were not formulated out of thin air. Although it is always dangerous to pin down cause and effect in a specific train of ideas it looks as if the plates are a consequence of the concept of 'transform faults', another brainchild of Professor John Tuzo Wilson of the University of Toronto, and one of the most brilliant insights in the history of oceanography.

Although the paper describing transform faults was not published until the spring of 1965, almost 18 months after Vine and Matthews had published their work linking magnetic lineations and reversal of the earth's field, their historic paper had created so little stir that Wilson had in fact worked out his theory without even knowing what they had done. Without the magnetics to help him he had arrived at an entirely new explanation of the behaviour of the faults between ridge sections by the direct 'pictorial' application of the idea of spreading in various configurations of cut and folded paper.

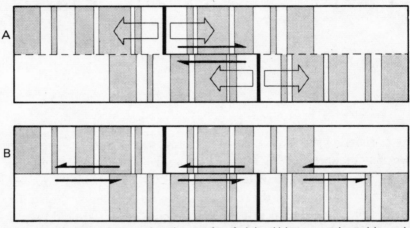

Comparison of relative movement along A, a transform fault in which new crustal material spreads symmetrically from two offset sections of a ridge, and B a transcurrent fault in which ridge sections were considered to be moving progressively apart from each other.

It had been assumed that the 'sliced' sections of a ridge, clearly visible in the charts, had originally been continuous but that faulting along fracture zones had occurred in such a way that sections of the ridge crest had moved away from each other so that they were now out of alignment. The fracture zones were therefore regarded in the jargon of geology as 'transcurrent' faults, in which the relative motion of the full length of the two sides was in opposite directions. Tuzo Wilson realised that if the sections of the ridge were maintaining their continuous production of new sea floor from their non-aligned positions, only the sections of the fault *between* the adjacent ridge crests would be in contrary motion to each other; the rest of the fault, on both sides of the crest, would share the *same* motion, so if there were to be any earthquakes on fault lines the place to expect them would be between the ridge crests.

This proposal satisfied the earlier observations that there seemed to be earthquakes only on some short faulted sections while the majority of the faults were inactive. Tuzo Wilson changed the name of the faults to 'transform faults', because they were 'transformed' into other features at the ridge crests, and he discovered another interesting thing about them – the two offset sections of ridge crest producing sea floor had not necessarily moved apart from each other at all, they could have been formed in an offset position in the first place. If so, the awesomely long faults, or fracture zones, observed in the Pacific, with their mile-high dramatic cliffs and chasms, were the result of slow and steady outpouring of separate narrow strips or plates of new sea floor from different starting positions. They were not, after all, the result of the cataclysmic wrenching apart of crustal rocks.

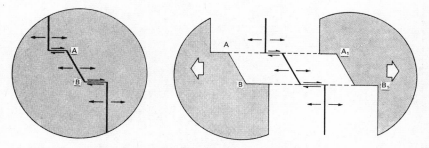

Diagram of continental separation from three sections of a spreading ridge. Places formerly close to-gether at A and B can be traced by following the direction of the transform faults to A and A₁, B and B₁.

Support for Tuzo Wilson's theory came from seismology. A new and improved generation of seismic instruments pinpointed the epicentres of the majority of all shallow earthquakes along the great ocean faults or fracture zones onto the section between ridge crests. Detailed analysis of a number of these earthquakes by Dr. Lynn Sykes of Lamont showed that the 'first motions' of the 'P' or pressure waves meant that the blocks along the fault between ridge crests *were* moving in opposite directions. This seismic evidence, collected to check on the validity of transform faults, was now in its own right a body of observation that is difficult to explain except in terms of sea floor spreading.

Tuzo Wilson himself pointed out that his transform faults meant that the offset sections of the ridge perpetuated the shape of the original break that had occurred when a ridge formed and rifted continental coasts apart. The faults, followed right across the ocean, for example across the South Atlantic, could connect places in South America and Africa originally joined together.

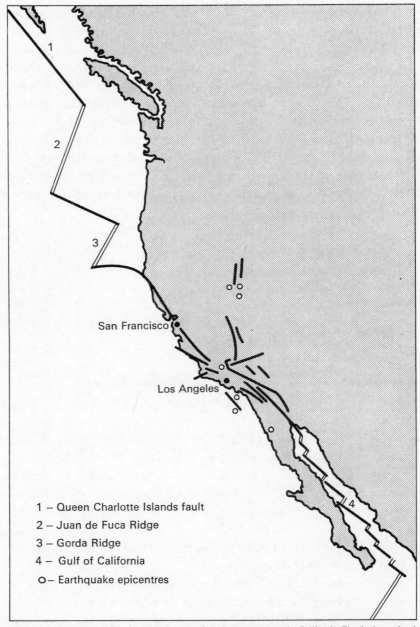

1 – Queen Charlotte Islands fault
2 – Juan de Fuca Ridge
3 – Gorda Ridge
4 – Gulf of California
O – Earthquake epicentres

Approximate positions of faults related to the San Andreas system in California. The faults on land conform to the direction of transform faults between short oceanic ridge sections in the Gulf of California and north-west of San Francisco.

We can see the early stages of the process in action in California. The Gulf of California appears to be a young ocean basin, probably about 10 million years old, opening up as the result of spreading from short offset sections of ridge. From the crest of the last north-westerly ridge section a great fault, the San Andreas fault, runs across both the sea floor and over-land to connect with the southern end of the Gorda Ridge, while another fault line connects the north end of the Gorda Ridge with the southern end of the Juan de Fuca Ridge. The whole of the connecting fault system can be regarded loosely as the San Andreas fault, the danger line that threatens the survival of San Francisco.

Being the section of fault between ridge crests the San Andreas is active and when the two sides move different ways in relation to each other, an earthquake is likely to result. The disaster of 1906 has left San Franciscans with a superstitious reluctance to use the word 'earthquake' at all but they are not the only people at risk. Los Angeles suffered severe damage from the movement of a minor branch of the fault in February 1971. Not un-naturally the San Andreas fault is the subject of intense and anxious study in the hope that the build-up of tension can be monitored sufficiently well by strain gauges and other instruments to give advance warning of an impending major earthquake. From these observations it seems that the whole fault does not act as a single entity, one side sliding past the other all along its length: instead some sections seem locked together, while others permit frequent small, safe movements to take place. Experimental work is now going on to see whether regular injection of fluids, even water, would lubricate the fault sufficiently to prevent the build-up of dangerous stresses.

Another young ocean basin is forming along the combined length of the Red Sea and the Gulf of Aden where offset sections of the spreading ridges are connected by transform faults which can be traced out as part of the original rifting process. Dr. Anthony Laughton of the National Institute of Oceanography in Britain has spent years piecing together the evidence that Ethiopia and the Arabian Peninsula were one piece of land only 10 or 12 million years ago, though the first signs of the impending break, the uplifting of a ridge-like structure, probably started many millions of years earlier.

Developing ideas of offset ridge crests and transform faults forced Tuzo Wilson to start talking about the sea floor as moving in independent strips or plates, each one spreading out from a short section of ridge and separated from the plates on either side of it by the boundaries established by faults. It is this notion of plates that has now been taken up and extended by other oceanographers into larger concepts of plate tectonics, very much concerned with the requirements and implications of spherical geometry. The people who virtually created the new theory were an American scientist, Dr. Jason Morgan, two Englishmen, Dr. Dan Mc-

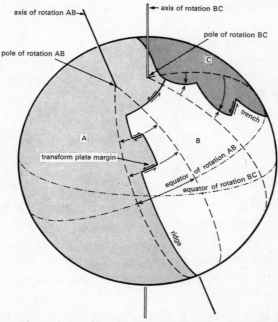

As rigid plates spread from mid-ocean ridges or are consumed at trenches, their movements over the globe must be rotations round 'poles of rotation'. The 'axis of rotation' is the line drawn from a pole through the centre of the earth. Each pair of plates involves a separate pole and axis of rotation, and a separate 'Equator'. Pole and axis AC, on the other side of the diagrammatic globe are not shown.

Kenzie and Dr. R. L. Parker; a Frenchman, Dr. Xavier le Pichon; and a three-man American team – Drs. Bryan Isacks, Jack Oliver and Lynn R. Sykes. Dr. Morgan started things off by applying Wilson's plates of moving crust to a spherical surface as a mosaic of fragments, each plate being bonded by a spreading ridge when new material is formed, a trench when material is destroyed (though this doesn't apply to every plate) and a fault separating the plate from other ones adjacent to it. In the eighteenth century the mathematician Leonhard Euler demonstrated the theorem that on a spherical surface the relative movement of any two blocks of material has to be a rotation around some point which is called a pole of relative rotation. The line passing from this point through the centre of the earth is called the axis of rotation. Applied to plates moving on the surface of the earth these become respectively the pole of spreading and the axis of spreading. So it is a geometric necessity that if plates of crustal material are sliding past each other along fracture zones, or transform faults, then the faults must lie along circles of latitude with respect to the pole of spreading. The rate of spreading at any particular point on a ridge must also be proportional to its perpendicular distance from the axis of spreading – in other words the spreading rate cannot be exactly the same all the way along a ridge on a spherical earth.

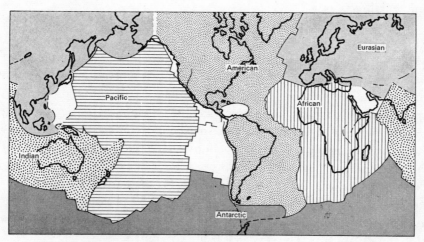

In its early stages the plate theory divided the earth's crust into six major plates and several minor ones, all bounded by ridges, trenches and transform faults.

Apart from providing geometrical explanations of such things as the varying widths of the spreading Atlantic and establishing that the geological relationships of oceanic features was far simpler than anything found on land, all this enabled Morgan to use the direction of transform faults, and the fact that they are bound to be more or less at right-angles to the ridge crest, to plot the various poles and axes of rotation of different ridges in different oceans. From changes in fault structure he was even able to deduce the position of spreading ridges in the distant past and to establish that the direction of spreading could change. Also, since opposing relative motions only occurred between small sections of adjacent plates, it was possible for Morgan to group whole blocks of narrow, 'mosaic' plates together into larger units. The result was a world-embracing pattern of six large plates, thousands of kilometres across, with a few smaller sub-plates thrown in. Movements of these plates can therefore be taken to account for the drift of continents inferred to be taking place at the present time.

McKenzie and Parker, working independently of Morgan, had come to some similar conclusions at almost the same time. They thought of plates as 'paving stones', essentially rigid blocks stretching from spreading ridges to an overthrusting or underthrusting boundary with other rigid blocks. Since all sea floor movement is outwards from ridge crests this means that the 'other rigid blocks' will be moving round the globe in a different direction.

Sitting in an office at Scripps, where they were working in the Institute of Geophysics and Planetary Physics, overlooking the Pacific, McKenzie says that they were suddenly struck with the realisation that if the continents round the Atlantic could be reassembled into an almost accurate fit, not only these, but the spreading sea floor that drove them apart, must be virtually undeformed and the whole mass of rock between trailing and leading edges must therefore be more or less rigid. With the Pacific stretching out below their windows, covering something over half the earth's surface, and with the old spreading ridge of the ocean running almost under the building, McKenzie and Parker started to try and calculate which way the Pacific floor was moving, by projecting the lines of its major transform faults and analysing the 'first motions' of earthquakes in the far distant trenches where it is being consumed.

The results made it seem certain that the whole North Pacific was moving as one paving stone or plate, between the west coast of North America and the Aleutian, Japan and Kuril trenches where an opposing Asiatic paving stone overthrust it. Although they satisfied themselves that a section of sea floor covering so vast an area could move as a single solid block they became very uneasy about the prevailing ideas of the mechanism driving the spreading system. With the pattern of large rigid paving stones plotted on the globe and with most of the crust-consuming trenches ranged round the edges of the Pacific, the upward and downward limbs of the kind of convection cells described by Holmes and Hess seem to be in very peculiar positions. 'It is', they wrote, 'difficult to believe that the convection cells which drive the motion are closely related to the boundaries of plates.'

In 1968, within a matter of weeks after publication by Morgan, Mc-Kenzie and Parker, a paper by Dr. Xavier le Pichon was in print showing how the rigid plates and their movement round their individual poles of

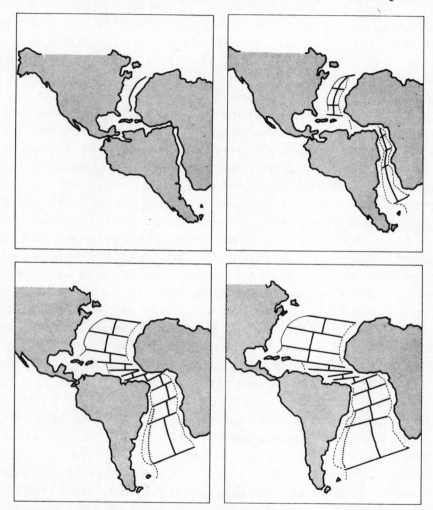

Reconstruction of four stages of the separation of the Atlantic Continents proposed by Le Pichon in 1968

rotation could be used to reconstruct the history of sea floor spreading all over the world throughout the Cenozoic era, which takes us back for about 65 million years.

Putting together the magnetic dating results, the known rates of spreading of different ridges at different latitudes and the directions of past plate movement 'fossilised' in the transform faults, Le Pichon's work amounted to a demonstration of how the spreading system could account for the position of all the present continents and especially for the progressive break-up of a supercontinental land mass centred on the Atlantic. He was able to draw diagrams of where the continents had got to at any particular time, when any particular anomaly would have been at the centre of the ridge. It all seemed too simple to be true and as a matter of fact it was, Le Pichon's original system of rotating plates on the surface of the globe turned out to be mathematically unworkable and has now been somewhat modified. When drilling and absolute dating techniques have completed the collection of the data it should nevertheless be possible to apply this general method of reconstruction even further back in time to the very beginning of all the present ocean basins. The clinching arguments to support much of the theoretical material was contained in the seismological evidence put together by Isacks, Oliver and Sykes in a paper published shortly after Le Pichon's reconstruction had appeared.

At first glance it is hard to see why the new plate theory is any advance on the broad outline of the sea floor spreading mechanism propounded 10 years earlier by Hess and Dietz, but in fact it has formalised some of the older ideas and enabled solutions to be found for some of the untidy topological and even physical problems raised by large-scale movements of the earth's crust. It has also emphasised the idea of the sea floor as a rigid and undeformed structure with enough strength to undergo its long journey in space and time across the globe, clarifying the imagery of what happens to it *after* it has been on its travels.

In Hess's picture of the sea floor mechanism we see the spreading ridges, marking the upward currents of mantle convection cells, producing new oceanic crust at a steady rate of around one centimetre a year. Where the ridges have opened up under a continent their spreading activity has rafted the broken sections apart, both of them keeping their symmetrical distance from the ridge, the results perfectly illustrated by the fit of the continents across the South Atlantic. Continental fragments drifted apart in this way eventually find themselves over the down-going current of the convection cell, marked by the great circum-Pacific belt of deep trenches. Too light to be carried down with the dense rocks of the ocean crust, the continents have settled above the descending limbs, their leading edges thickened and distorted into mountains, overriding the trenches and gradually forcing them to migrate backwards into the shrinking Pacific. Ocean floor uncluttered by continents is also rafted outwards from spreading ridges but is destroyed at the trenches, with earthquakes marking its return to the mantle at a steep 45° angle of descent. The clearest example of this operation is the moving floor of the Pacific, once presumed to have spread outwards from what Hess and Menard thought was the decayed 'Darwin Rise' in the western part of the ocean, but now spreading from a younger feature, the East Pacific Rise, neither of which lay in the middle of the basin because they had not opened under a continent in the first place.

Hess suggested that the cellular flow marked by the ridges and trenches would make the earth look rather like a tennis ball stitched in dumb-bell sections but from his papers it is also possible to imagine that he had in

SCHEMATIC BLOCK OF OCEAN CRUST.

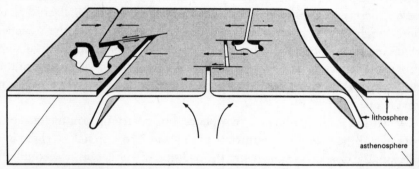

lithosphere

asthenosphere

Rigid plates of lithosphere bounded by ridges, trenches and transform faults, slide over the plastic or molten layer of the asthenosphere which in turn lies above the mesosphere.

mind one semi-complete shell or cap of crustal material, spreading outward from the ridges of the Atlantic, the Indian Ocean and the Antarctic, slowly overriding and reducing in area another cap of crust represented by the Pacific.

In contrast to all this the picture presented by the new plate theory sees the earth's crust as far more fragmentary and far less closely linked to a system of convection cells in the mantle, although heat supplied from the interior of the earth is still obviously the major factor in all crustal movement. The crust still forms curved caps fitting over the spherical earth, probably as much as 100 kilometres thick and making up a layer of material called the lithosphere (sometimes called the tectosphere) which is divided into a number of rigid plates sliding across the globe on a deeper layer of plastic or molten material called the asthenosphere. Instead of being governed directly by the behaviour of the convecting mantle cells the continents are now thought of as 'passengers' carried by the moving plates.

At their trailing edges all plates are very much alike; their differences depend on the behaviour of their leading edges. A simple plate will be formed from one side of a spreading ridge, new material welling up from the mantle to fill the rift as the crust is pulled apart so that the trailing edge of the plate is constantly renewed. At their leading edges some plates, like the Pacific plates, consisting entirely of dense basaltic oceanic crust and sediments, descend into trenches and are consumed. Other plates, like the long Western Atlantic or 'American' plate, have continents on their leading edges and since the light granitic continental material cannot descend into trenches, the plate overrides any plate or plates it meets travelling in the opposite direction, in this case the plate of the Eastern Pacific. It looks as if the northern end of the American plate overrode the West Pacific plate some hundreds of millions of years ago, covering not only the trenches which may have bounded it but even covering the main northern section of its spreading ridge as well. The narrow section of plate to the east of the East Pacific Rise, with its trenches still active, survives as 'evidence' of what happened.

When the leading edges of two plates, both carrying continents, confront each other, neither continent can descend into a trench so there is a collision, the classic example being the collision of India with the southern coast of Asia which has resulted in the raising up of the Himalayas.

The movements of India and the spreading pattern of the floor of the Indian Ocean turn out to be exceedingly complex but put at its simplest India is presumed to have been driven under Asia by the spreading of the Carlsberg Ridge. The Indian Ocean in its present form is therefore a young ocean; while the ocean that must have separated Asia from India in its earlier position, the eastern end of the Tethys Ocean, is presumed to have been entirely consumed by a trench. This trench has now vanished under the colliding continents but its activity is still shown in the earthquakes that occur under the whole Himalayan region, stretching westwards to the Alps. In fact this trench is thought to be part of a still active system which would have been a far more obvious feature if there had been any oceanographers around to observe it before the ancient Tethys Ocean started to close.

The plate theory allows itself to undergo constant modification as new ideas are developed and it has happily accommodated the disappearance of ancient oceans and the opening of new ones as the plates moved into their present observed positions.

THE TETHYS OCEAN

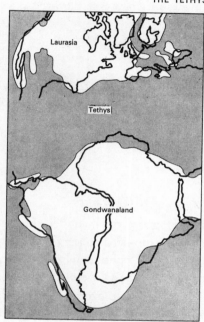 

Two alternative positions for the ancient Tethys Ocean depend on whether or not Laurasia and Gondwanaland were once joined together.

Playing with plates is now the major preoccupation of oceanographers and geophysicists. To a frivolous onlooker it is a new game (it has already been called a kind of spherical chess), in which pieces can be moved across the globe to provide solutions for all kinds of awkward problems. Ingenious players can invent new sub-plates, fragments left over from older stages in the system, if they run into intractable facts – like for instance the seismic activity of the Caribbean – and there is plenty of room for manoeuvre in explaining the many stretches of aseismic, inactive ridges that are lying about in different stretches of the ocean. When certain plate movements proved really baffling, McKenzie and Morgan recently came to the rescue with a whole battery of geometrical solutions. They allow transform faults to be created when two or more plates meet or when related trenches are required to consume oceanic plates moving in opposite directions.

Reassurance about the validity of the game in disposing of sea floor down trenches has also been provided by new work from Lamont seismologists Sykes and Isacks, who have elegantly demonstrated that seismic waves from earthquakes travel up an inclined plane below the trenches at fast speeds similar to those at which they move through the dense rocks of the sea floor, reinforcing the theory that the lithosphere turns down and is consumed at these sites.

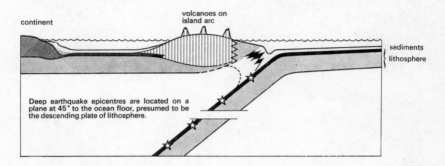

Deep earthquake epicentres are located on a plane at 45° to the ocean floor, presumed to be the descending plate of lithosphere.

Over the last decade, as the great new geophysical model emerged from the work of oceanographers, all but the die-hard conservatives have been converted to its acceptance, and with it to continental drift. Some enthusiasts like Tuzo Wilson and Vine and Matthews were ahead of the field, with Blackett and Runcorn leading a separate group entering the 'faith' through palaeomagnetism. Irving's work on wandering magnetic poles led Bullard to 'testify' at the first International Oceanographic Congress held in the United Nations building in New York in 1959 and it was the subsequent Bullard, Everett and Smith work on the fit of the Atlantic continents that finally convinced Patrick Hurley, Professor of Geology at the Massachusetts Institute of Technology.

Today almost the only oceanographers and geophysicists who are wholly opposed to drift, and spreading, are certain distinguished Russians, notably Professor V. V. Beloussov of the Russian Academy of Sciences in Moscow. Working from the home base of a huge continental land mass they insist that all the geophysical problems can be explained in terms of vertical movements of crustal blocks and that even ocean basins can be explained as foundered continent. On their side it must be admitted there still stands a considerable body of old-school geologists who, working within the confines of small areas analysed in detail, have failed to feel the relevance of the new ideas. Curiously enough the geologists of the competitive commercial industries like mines and oil companies, who stand to profit most from advances in geological thinking, have not, with some honourable exceptions, found it necessary to adapt their detailed findings to the new geophysical framework. There are signs though that the newest discoveries of oil and minerals in the sea, beyond the limits of the exploitable continental shelves, mean that even the most conservative companies are rethinking their broad strategy in terms of drift and that even the Russians have their 'defectors'.

7

Drift Everlasting

Plates are likely to keep quite a number of the world's oceanographers busy for several more years and long before they run out of
work the elusive clues to the driving mechanism for the whole system will
probably have turned up. They may indeed already exist, only waiting for
sharp eyes to discover an unsuspected new pattern in the data or they may
have to wait until new knowledge about the rest of the solar system
allows us to reconstruct the thermal history of the earth.

Meanwhile the superficial pattern of plate movement is sufficiently
complex and philosophically satisfying to be in danger of lulling even
some of the most far-seeing oceanographers into the belief that, with the
establishment of continental drift as an article of faith, we are within reach
of solution of all the major questions of geology, providing a matrix
within which all minor puzzles can be solved. The chances are that they
are wrong, settling down to the relaxing task of tidying up the loose ends
at exactly the moment when the next stage of the action is about to begin.

The mere process of assembling material for these chapters raised
questions that are badly in need of answers; questions far too large to be
confined within the scope of any 'tidying up'. There are, for example,
signs that certain areas far inland in the continents have at some time in
their history faced into deep sea water; that as well as oceans changing

position some oceans that existed in the distant past have vanished. The break up of the supercontinents of Gondwanaland and Laurasia may explain the shrinking of the Tethys Ocean but not the older signs of the sea.

Then there is the question of what was happening to the rest of the earth's surface when the continents were all bunched together. Plotted on a globe the rest of the surface turns out to be the floor of one great ocean, a kind of super-Pacific, 'Panthalassa' as Wegener called it in 1915, in which all lesser oceans can be considered as bays around its margins. Was this the primeval ocean and if so did it perpetually renew itself by spreading, either from the ridges surviving today or from others now lost in the devouring trenches? If so, what kind of plate pattern was operating then? Most insistent of all was the question of why the original supercontinental land mass should have split up at all. What happened only 200 or 300 million years ago to start the 'Great Drift', when the continents had stayed in one, or at most two, pieces for 4,500 million years or more? What was the sudden signal for so drastic a change and why can we find no traces of its cause?

Sometimes these questions received a sympathetic response from other people disturbed by the awkward loopholes and inconsistencies in the broad general theory. Others were so absorbed in the details of the work in hand that they had neither the time nor the inclination to consider the wider issues. Still others, though only a few of them, believed with an almost religious fervour that to stray from things that could be observed and measured into such large and almost abstract issues was in some way a breach of faith with the very principles of science. To make large coherent patterns out of the results of many small and scattered scientific advances was the shady business of the populariser, only a trifle higher in the scale of respectability than the black art itself.

From one person and another the awkward questions provoked a curious ragbag of ideas, some of them original, some of them left over like fossils from earlier stages along the route of development of the general concept of the moving sea floor. The idea of the expanding earth is still cherished in some quiet academic corners, surviving side by side with the picture of mountains raised by crustal shortening on a cooling, shrinking earth.

One idea with a long history that still turned up occasionally to confuse the mechanism of sea floor spreading was the suggestion made in 1878 by George Darwin (the son of Charles Darwin), that at some molten stage in the earth's history the moon was flung or pulled out of it, like a huge droplet, carrying away a large section equivalent to two-thirds of the cooling crust and with it the light materials that would have formed a 'continental' shell over the whole surface. Now, according to this notion, all that remains of that early crustal rock is the material that forms the continents while the great scar left by this tremendous event became the Pacific Ocean. Darwin's collaborator, Osmond Fisher, even went so far in 1882 as to suggest that the continents, after breaking up when the moon departed, were likely to readjust themselves to more symmetrical positions on the globe.

Darwin's proposal about the moon's origin is distinctly back in fashion at the moment because the alternative, that the moon was captured by the earth, demands very complicated astronomical circumstances. As our moon is after all only one of thirty-odd moons in the solar system we need to discover a general moon-manufacturing process and elaborate captures on such a scale seem less plausible than some form of ejection or consolidation of satellites by their mother-planets. Yet even if the Darwin theory is right in explaining the origin of the moon the Pacific is not, as everyone thought in 1878, a single, undisturbed, primeval ocean but in fact has the same kind of recent geological history as any other ocean basin, namely slow spreading from a ridge axis.

The stock answers to questions about the long cohesion and sudden break-up of the supercontinent were always based on a picture of changing convection cells within the earth's hot mantle, a picture that owes its popularity to Professor S. K. Runcorn of the University of Newcastle. He made some calculations based on the proposal that a succession of cells formed as the nickel-iron core of the earth grew larger in size, and destroyed the primitive single cell pattern. First the original supercontinent broke up and as the cells increased in number and changed their position the disturbances were marked in successive spasms of mountain-building known to have taken place at definite intervals 200 million, 1,000 million, 1,800 million and 2,600 million years ago.

Several geophysicists seemed to believe that one or two supercontinents, implied by the fit of the continental margins and the sea floor dating, had been formed before the Great Drift began out of fragments brought together in earlier drift episodes, but they did not have any clear idea how or why such a centralised collection of collisions could have been made or what could have happened to the oceans that formerly separated them. 'You can't ask me to believe', said one, 'that an ocean the size of the Atlantic or the Pacific could just vanish without trace.'

One man did believe just this, Professor Tuzo Wilson of the University of Toronto, the indefatigable theoretician and synthesiser of new pieces of information, and the author of the brilliant transform fault theory. For some years he has been moving towards the conclusion that oceans have a life cycle of birth, growth, maturity, decline and death, their disappearance marked by a 'geosuture', a line of junction between two different rock masses such as the Indus line in the Himalayas, to show where two continents have collided. He has also suggested that the present Atlantic Ocean was preceded by an earlier proto-Atlantic which in closing caused the uplift of a great continuous mountain range now broken into the Appalachian mountains along the eastern seaboard of North America and the 'Caledonides' mountains of Scotland and Scandinavia.

Such a proto-Atlantic, which in closing brought North America into collision with Europe and Africa, led him to believe that the present continents are a mosaic of fragments brought together and broken apart over and over again by the life and death of oceans. The eastern United States has retained fragments of African and European rock left behind by the fracture of the present spreading ridge, while on the western seaboard of North America fragments of Asiatic rock may be embedded in what is now Nevada, British Columbia and California. Even the vegetation, he says, reflects these earlier lines of junction, the trees and plants of western North America having a distinctly oriental ancestry while those east of the Rockies are European.

These imaginative suggestions had a lot of informal support from the kind of land geologists who were trying to make sense of large-scale effects and structures. These people tended to feel that if the geophysicists could only come up with one pattern of drift, starting 200 or 300 million years ago, they were not solving many of the geologists' problems. In spite of this, the balance of opinion seems to have swung recently in favour of only one Great Drift episode. After an elaborate collation of rock dating from all over the world, Professor Patrick Hurley of M.I.T. and another geologist, John R. Rand, concluded that the oldest rocks in the world, dating from between 1,700 and 3,000 million years ago formed such a coherent pattern that they looked very much as if they had always made up two primary continental nuclei that had not been either scattered or

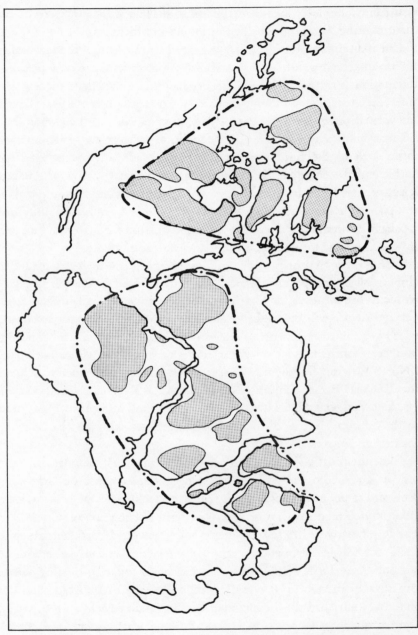

Coherent grouping of oldest continental rocks led Hurley and Rand to believe that only one **Great Drift** episode took place. Hatched areas represent rocks older than 1,700 million years.

brought together by any drift motions before the present Great Drift began. These old nuclei seemed to them to make up almost a third of the whole continental area and to have grown to their present size by the gradual addition of more or less concentric rings of younger rock. Some of the new material was in the form of mountain chains similar to those found when sea floor moves under the edge of a continent, so they thought the sea floor motions could have existed before the Great Drift began, though they did not offer any suggestions about its pattern.

It seemed ungrateful to go on asking for better answers when oceanography was producing so many exciting results which deserved applause and congratulation. From the point of view of oceanographers themselves it probably seemed unprofitable as well. Their business is the ocean as it is now and the traces it contains of its history over the last 200 or so million years. Further back than that the evidence is missing, probably lost forever in the maw of the trenches and ideas about the more ancient past tend to be classed as mere speculation, an activity synonomous in many scientific circles with immorality. Scientists are still sensitive to the jibe of Mark Twain, 'What I like about science is that one gets such enormous dividends of speculation for so small an investment of fact.' Pinned against a wall and argued with, however, even an old-fashioned geologist will now concede that there is a place for speculation in determining which of many alternative lines of enquiry is likely to get results. In this case speculation led straight to land geology, the stronghold of the unconverted; for if the sea floor is comparatively young the rocks of the continents are 10 times older, some have been dated at more than 3,000 million years, and if any evidence remains of the plate patterns of the distant past it will be found in them or not at all.

It was this broad background of uncertainty that made all the more startling a geological paper published in 1969 by Dr. J. F. Dewey of Cambridge University: indeed it was startling to find that geology still has the power to startle.

Dr. Dewey has made a detailed geological analysis of the rock sequences of the Appalachians in North America and the 'Caledonides' in Scotland and Scandinavia, theoretically reassembled into a continuous range. The structure of the range is complex. It is commonly recognised by geologists that it can be divided into distinct zones; in places there is a clear northern zone in which the rocks have been metamorphosed or changed by heat, and a clear southern zone where they have not. In his paper on the evolution of the range Dr. Dewey produced a convincing model of the various stages of closure of an earlier Atlantic Ocean by the consumption of a plate by trench systems along both its margins. He showed how different kinds of rocks were produced at each stage in the process and how the descent of the plate had deformed and uplifted the continental margins, finally bringing them together along a suture line which marked the last remnant of the ocean floor.

This was going back beyond the start of the Great Drift with a vengeance and as it seemed to be the first piece of authoritatively documented work in that direction, it was worth tracking Dr. Dewey down and asking him just how far along that road his ideas had taken him. Active geologists who climb mountains to get their material are nearly as difficult to meet as oceanographers who go off to sea for weeks at a time in research ships, but he finally made an appearance in an office at Lamont where he was based for a short summer research fellowship.

John Dewey admitted to having been attracted by the idea of continental drift when he first met it at the age of 18 at a lecture by Professor Lester King in London and was now a whole-hearted convert to plate tectonics and a Cambridge friend of Dan McKenzie, one of its principal exponents. Once the plate idea had taken hold of him he saw geology within its framework and concentrated on looking for the geological signs of what might be going on at different plate margins. His work on the Appalachians had led to another paper, still in the press when we met, on the general principles of how an Atlantic type of non-volcanic continental margin with a long accumulation of sediments at the continental shelf, from erosion of the land, could be converted into an Andean type of margin consisting of a volcanic mountain range in which the old sediments were deformed and uplifted by the opening of trenches in the ocean.

This conclusion brought John Dewey by his own route and with a wealth of detailed geological evidence in hand to very much the same position reached at just about the same time but on broader and more general lines by Tuzo Wilson.

Dewey also believed that oceans had a life cycle; opening from a spreading ridge, maturing as wide basins with a heavy marginal accumulation of sediments, declining with the formation of 'eating' trenches and finally dying when the whole basin, including the spreading ridge itself, had returned to the mantle. His studies also appeared to confirm the idea that the Urals, one of the few highish mountain ranges in the world to be found far from a continental margin, were the memorial to a fairly recently closed ocean, with a suture line marking the junction of the previously separated continents.

It was probably a lucky chance that John Dewey had been so hard to find, for it meant that a detailed explanation of his work came at the end of a period of several weeks spent in the U.S. and Canada talking to oceanographers about their latest findings, checking up on the latest research results, discussing their implications and the new problems they raised, hearing at first hand the unpublished, unpublishable speculations and conjectures, the unguarded opinions on other people's work. Talking to a plate tectonics-minded geologist after all this meant there was so much to discuss that the talk went on through most of the day, into dinner and on into the night, among a growing pile of maps and diagrams and charts of the ocean floor.

Just looking for any length of time at the map of the world with the spreading ridges and trenches clearly marked can be as good a way as any of inducing a mood of 'creative speculation'. Looking at it in the particular circumstances of the discussion that evening seemed to produce more than the usual rush of ideas to the head. We looked again, with the usual wonder, at the huge area of expanding plates embraced by the Americas, the Atlantic, Africa and the western Indian Ocean, and at the sweeping double line of trenches that seem so nearly to encircle the Pacific, like the strings of a purse, or the lines of a seine-net. Shorten the string and the purse or the net closes: spread the Atlantic, the Arctic and the Antarctic Oceans so that the continents on the leading edges of the plates move farther into the Pacific, and the purse-strings of the trenches must surely shorten. The implication was clear: if the Pacific is closing, the trenches, to survive, must retreat into the ocean basin ahead of the continental margins and to do so they must somehow 'unroll backwards', their point of descent constantly changing.

It was in fact an idea that had occurred to Hess back in 1962 but in this form it seemed worth the investment of a little time, at least to check the 'first motions' of the seismic analyses at the trenches to see if it stood up to examination. If it did, stated in formal terms, the retreat of trenches before advancing continental margins could be considered as a control mechanism for the rate of plate consumption. Back in London the implications of this long conversation with John Dewey refused to lie down tidily in a notebook and a tape recorder. Instead they started a kind of obsessional mental itch, returning again and again to the surface of the mind until it became impossible to think anything else except oceans; oceans opening as spreading ridges split continents and drove the pieces apart before them; oceans closing under the attack of 'eating' trenches, continents moving inexorably together to form larger continents, continents moving in the *wrong* direction to conform to the pattern of the Great Drift. Somewhere in all this movement there had to be a system and suddenly, quite unexpectedly, it seemed as if there was.

What emerged, after a great deal of imaginative effort to sustain the number of variables involved, and a great deal of fooling around with maps and visual aids in the form of paper cut-outs of the continents stuck on a globe, was nothing less than a new model for plate movement as part of a dynamic continuous process. It carried with it the inescapable but startling conclusion that the continuous creation and destruction of sea floor plates was also a mechanism that could account for the formation and entire growth of the continents themselves. Instead of being largely created in one catastrophic episode continents had evolved from small beginnings. Instead of their marginal mountains being formed by sudden changes in the pattern of mantle convection cells they had been built in a logical and more or less predictable sequence whenever an ocean started to die.

Having arrived at this presumptuous notion it seemed worth telephoning John Dewey at Lamont to see if he laughed. He didn't, but a transatlantic telephone call is not the ideal medium for speculations on geophysics, so the next stage of development of the idea took place less extravagantly by mail. After many months of detailed work by John Dewey, and great exchanges of correspondence followed eventually by meetings in England, the new version of the plate game took shape and was published, in February 1970, as a scientific paper 'Plate Tectonics, Orogeny and Continental Growth' (Dewey and Horsfield) to take its chance in the wide arena of oceanography, geophysics and geology.

The real starting point for the new model is the idea that the descending plate margins *cause* the production of volcanic mountains of two kinds, island arcs in the open ocean and coastal mountain belts like the Andes where the plates descend close to a continent. It is also important that these volcanoes produce lavas of the continental 'andesitic' type, instead of just the basalts found in places like Hawaii and other typical 'oceanic' islands.

Behind an island arc, or a chain of linked arcs, muds and sands eroded from the rocks of the continental mainland fill the intervening strip of sea with sediments that are slowly turned by time, and heating from the descending plate, into layers of sedimentary rock which effectively extend the continental margin to embrace the new island arcs. They in turn will be further extended by the addition of some marine sediments scraped from the surface of the plate as it descends.

The coastal mountains, which geologists call the 'cordilleran' type after the Cordilleras in western South America and of which the Andes are typical, are less effective than island arcs as manufacturers of new continent. The mountains themselves are made of several different kinds of rock. Close to the site of the trench there are wedges of oceanic crust and sediments scraped off the plate surface, all apparently affected by a regime of low pressure and high temperatures. Further away from the trenches are found the sediments of the continental rise, laid down before the opening of the trench system, and now thrust up and altered or 'metamorphosed' by heating. Erupting through them are granites and the lavas of 'continental-type' volcanoes. It is these sequences of events which account for the formation of the thick sedimentary layers originally ascribed to the action of geosynclines. (Anyone astounded by this proposal may care to recall that as long ago as 1940 Harry H. Hess said, 'Perhaps no greater misconception exists in geology than that geosynclines localise mountain building.')

When these mountains erode under the combined effects of frost, wind and running water, new sediments are produced which may accumulate on the coastal strip of oceanic crust between the continent and the trench, though this is usually too narrow to allow the continent to extend very much horizontally at this stage. By contrast oceans which do not have trenches, for instance the greater part of the Atlantic, have plenty of room for sediments eroded from the continental rocks to be laid down at the ocean margins, forming the thick layers of the continental rises.

These processes fall into two categories: true continental accretion in which new material is added to the continent by volcanic eruption or by being scraped from the ocean bed, both of which add volume to the continent, and pseudo-accretion which is the result of erosion of the new material into sediments which add no volume but only spread the material over a wide area. It is by the combination of both processes that the continents grow.

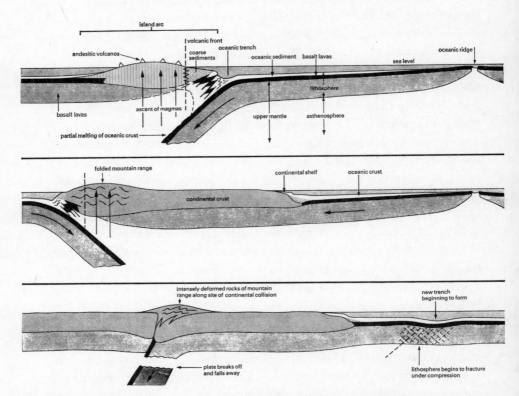

Three types of plate boundaries: (*Top*) Two plates of oceanic crust are moving in opposite directions. One is under-thrusting the other and an arc of volcanic islands has erupted in the upper plate. Example: the island arcs and trenches of the Western Pacific. (*Middle*) This time only one plate is oceanic; the other is continental. The oceanic crust is under-thrusting the continental one and a cordilleran range of mountains has formed at the edge of the continent. Example: the Andes. (*Bottom*) Here the edges of both plates are continental and the two continents are in collision and a complete mountain range marks the junction. Example: the Himalayas where the collision of India and Asia still seems to be in progress.

The present pattern of the moving crustal plates shows some oceans that are simply spreading from mid-ocean ridges and are therefore clearly opening while some oceans although they have spreading ridges, have consuming trenches as well. Now unless the circumference of the earth is expanding after all the production and consumption of lithosphere in the sea floor spreading process must balance. So it must be presumed that the trenches are consuming not only the output of ridges in their own oceans but of all the other ridges also. They must in fact be consuming faster than their 'own' ridges are producing, and therefore their oceans must be closing. This means that not only the Pacific is closing but the eastern part of the Indian Ocean together with the Tasman Sea and the Mediterranean, all of them combining to destroy the entire global production of new sea floor.

It is not hard to imagine the process carried forward in time to the point where some of the closing oceans close, their opposing continental margins brought into collision, their spreading ridges inactivated and swallowed by the trenches, until finally the trenches themselves stop work. Meanwhile the opening oceans, with their spreading ridges unimpaired, are still steadily opening and producing new sea floor, only now they are deprived of the convenient services of the lost trenches to dispose of it. To solve the geometrical problem new trenches are required and the only place they can form is in an opening ocean. Fractures break the lithosphere, either close to a continental margin or further offshore, the ocean floor sinks downwards and the new trenches are born.

The idea that trenches will open in an ocean when its 'lithosphere disposal problem' becomes acute is rather appealing. It gives those wide, impersonal, even hostile, oceans a certain quality of human frailty, like householders driven to desperate action by a strike of garbage collectors. Scientific caution, however, insists on admitting the less entertaining possibility that trenches simply appear when an ocean floor that has been opening for hundreds of millions of years and, steadily cooling and dropping below the level of its ridge, sags under the weight of sediments gradually collected in deep piles along its distant margins. Whatever the precise starting mechanism turns out to be, and it will probably remain in some doubt since none of us is likely to witness the birth of a trench, the consequence is that the ocean in which the new trench or trench system is formed, immediately becomes, potentially at least, a closing ocean.

It is possible, for instance, that the closing of that part of the Tethys which lay between India and Asia before their collision will eventually require a new trench to form on the Indian Ocean to the south-west of India to take up the lithosphere being produced by the Carlsberg Ridge. Or if, in the course of a few hundred million years, the Pacific finally closes up to make Asia, North America and South America into one supercontinent, the Atlantic would be forced to open trenches either along one or both of its margins. Then, even though the Mid-Atlantic Ridge was still adding new material at the trailing edges of its two Atlantic plates, the ocean would start to close. Of course the lithosphere production and consumption rate of the ocean might exactly balance and it would then remain the same size but it seems more likely that the trench formation would be the signal for a new rift to form and start opening up a new ocean, a 'neo-Pacific', either along an earlier line of weakness at the site of the line of closure of the old Pacific or in some new position that split the 'Asiamerican' supercontinent apart.

The total closure of the present Pacific may seem fanciful but it could well be that this kind of cyclic opening and closing of oceans by plate action is exactly what has happened in the past for the greater part of geological time. If so, it is worth mentioning that it is necessary to make a slight modification to the plate theory to allow continents to change or even reverse their direction of movement as oceans close and other oceans open. The current theory is that continents are carried as superficial passengers on plates and therefore by implication 'belong' to them. Yet in the Pacific-Atlantic example we have just seen, the North and South American continents carried by the present Atlantic plate would be separated from it by the opening of new trenches along the eastern United States or eastern South America let us say, somewhere perhaps between Venezuela and the Argentine, making a new plate boundary. In their next cycle of movement the continents would therefore become the 'property' or 'temporary guests' of another plate and their direction of movement would depend upon the axis of its own spreading ridge.

Without embarking on mental gymnastics this demands a change in the 'philosophy of plates'. Until now we have been asked to visualise the confrontation of two ragged plate margins, one oceanic and the other carrying a continent on its leading edge, moving in opposite directions and 'forcing the issue' so that the denser oceanic margin is 'obliged' to under-thrust the more buoyant continental one. Instead we should see the underthrusting of the oceanic plate as the solution to an earlier 'lithosphere disposal problem' caused by its own spreading. The formation of trenches, the descent of oceanic crust, the creation in fact of a new plate boundary *allow* the continental margin to become a plate leading edge.

Armed with these considerations and with such evidence as the closure of the proto-Atlantic and the Tethys, and the continental collision at the Urals it is now possible to use our new model to look backwards into time for the history of plates before the present Great Drift began. The clue to the vanished oceans of the distant past will be the mountain chains created at what were the continental margins, near the site of lithosphere con-sumption and their directional axis will tell us very roughly the direction of plate movement. It should be possible, given time and money and insatiable geological curiosity, to unravel the complex accretion processes of the continents, working backwards from one leading edge to another, each one older than the last, each chain of mountains more eroded and ground away by time, until only the original rocks of the continent are left, the rocks of the continental nucleus. In fact the rough dating of the continental rocks does reveal just this kind of broad pattern conforming to a complex sequence of accretion on different leading edges with the oldest rocks of all, the 'shield areas' and 'cratons' more or less surrounded by newer material.

How far back can the unravelling be taken? Among the oldest rocks in the world are those of the Rhodesian Craton in Africa and there is some evidence to support the idea that in this sequence the typical lavas of island arc volcanoes are underlain by what could be sections of sea floor laid down more than 3,000 million years ago. The same may be true of the Canadian shield where there are signs that the earliest of island arcs may have cut off areas of even older sea floor complete with basalt pillow lavas that must have been erupted under water and were raised up by the later isostatic uplift of the whole continent.

These possibilities encourage the imagination to return to a time when no continents existed at all. The ocean, if indeed it was there, can only have been a relatively shallow covering of water over the earth because so much of the present volume of the oceans has been produced by the processes of rock formation themselves. Driven by forces we still do not understand the most ancient rocks that covered the whole surface of the earth may have split into plates and grown and moved and underthrust each other for the first time to produce the first volcanic islands of original continental

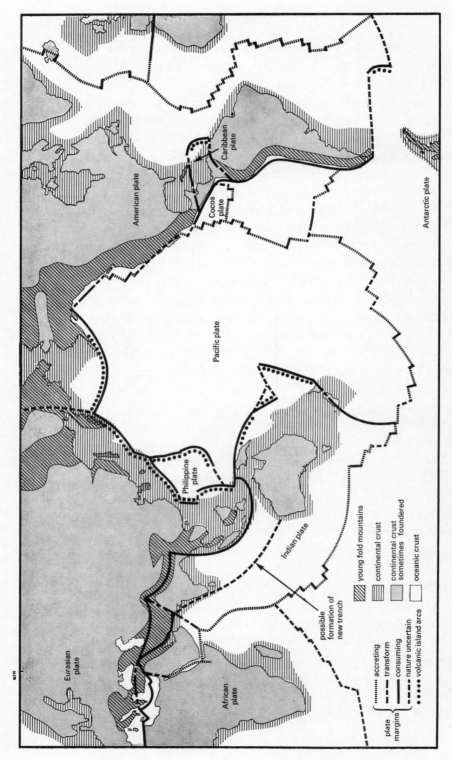

Boundaries between plates are either spreading ridges at which new lithosphere is made (accreting), trenches in which it is destroyed (consuming), or faults along which plates slide past each other (transform). Island arcs and young fold mountains mark the sites of current or recent lithosphere consumption along detectable or presumably overrun trenches of closing oceans. A new trench may eventually be 'required' in the North-west Indian Ocean following the collision of India and Asia.

American plate

Caribbean plate

Cocos plate

Antarctic plate

Pacific plate

Philippine plate

Indian plate

Eurasian plate

African plate

possible formation of new trench

young fold mountains

continental crust

continental crust sometimes foundered

oceanic crust

volcanic island arcs

plate margins

accreting

transform

consuming

nature uncertain

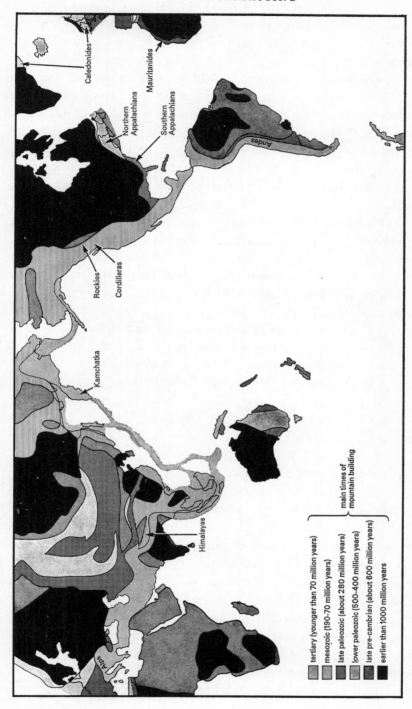

Pattern made by dating of continental rocks supports the idea that continents have grown outwards from ancient nuclei by the addition of new material to successive leading edges during many different plate movements as oceans opened and closed.

material, too buoyant ever afterwards to return to the mantle of the earth.

We have now entered the region where the lack of a proper physical explanation of the cause of plate rifting turns all argument into a matter of 'maybe' and 'probably' and 'perhaps'. The fact that oceans both open and close makes it difficult to believe in any kind of one-for-one relationship between convection cells in the mantle and the surface pattern of ridges and trenches. Certainly it has been well established that at high temperatures plastic creep could, and probably does, allow convective overturning of some kind to take place in the apparently 'solid' mantle material, although a satisfactory model for its real motions in the earth has not been produced.

Satellites have reported evidence of density anomalies on the earth's crust that have been interpreted as having some relationship to the position of ridges and trenches and there is in effect a slow cellular circulation of hot material rising through the ocean rifts, cooling as it moves outward across the ocean basin and returning down cold to the mantle in the trenches. But the 'cells' made by the present ridges and trenches have very strange shapes and it is hard to visualise them expanding and contracting to keep pace with the changing surface configuration of continents and oceans. Although it may seem a philosophical nicety it may be profitable to consider the trenches as a 'lithosphere disposal service' and the whole cellular motion of the surface as an effect of rifting rather than its cause. We should then be forced to look for some other kind of cellular convection pattern, or indeed something else altogether, to account for the rifting itself.

When the continents first began to form and grow the oceanic lithosphere plates may have been far more numerous and possibly even thinner than those of today. As ridges and trenches played out their cycles of production and consumption they would slowly have increased the island arc complexes by the eruption of basalts and andesites and their acidic derivatives, gradually amalgamating them into larger continental nuclei. As erosion products from the new islands and their mountains had a chance to accumulate without disturbance at the margins of opening oceans the cordilleran type of mountain building would have become possible, getting steadily more important as the system developed. As the continents grew larger their patterns of movement must have become more complex, the shapes of the land masses created by frequent collision becoming more ungainly, so that even if, at the earlier stages in the process, the new rifts always opened along the sutures, the lines of weakness left by previous collisions, without disrupting continents at all, this would gradually have become impossible and established blocks would have been broken up to make new formations half-way across the world.

Until the mystery is revealed by new work on the physics of the interior of the earth or even of its surface, we do not know how plate movement was begun, nor how it may end. The proposed pattern of continental movement and growth, carried far enough into the future, could cover the whole surface of the globe with continental rocks forming a characteristically buoyant crust 20 kilometres thick, usurping the whole domain of the oceans, and leaving no space for new sea floor. Apart from the fact that this would also leave nowhere for the water to go except up and over large tracts of continent, how would the plates behave then? Would they stop or would they enter a new mode of behaviour?

Before such a possibility can be put to the test the whole solar system may have come to an end so we can only reasonably argue about shorter-term possibilities. From the present pattern of plate movement it is possible to believe that nearly all the continents could be assembled into a single supercontinental group in the position now occupied by the Pacific, leaving a super-Atlantic to sustain the world's 'bilateral asymmetry'. In fact, until the Pacific has closed it is hard to see why the expansion of the currently opening oceans should be checked. It is also quite possible, though by no means certain, that a single supercontinental land mass *was* temporarily formed somewhere between 200 and 300 million years ago when the Pacific was in its opening phase, though Wegener's Pangaea, which doesn't allow room for the Tethys, is not quite the appropriate configuration. No doubt the current work will allow a better model to be compiled before long.

Such large continental assemblies may not, however, be necessary to the working of the system. It is possible that the opening and closing of oceans causes a continuous process of continental collision and fragmentation in groups of different sizes and that continents are more or less constantly on the move. Until the ocean dating has been completed and indeed until an absolute time-scale for all oceans has been established by the results of drilling, more comprehensive rock dating and other studies, there is no way to be sure what the recent movements of all the continents have been. Very little work has been done for example on the many continental fragments that may have gone into the building of Asia, nor is

THE LIFE AND DEATH OF THREE OCEANS

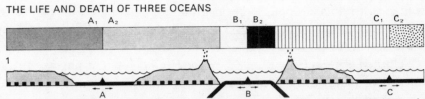

1   Three oceans A, B and C, each with a mid-ocean ridge from the axis of which two plates spread outwards. (Tinted areas above diagram indicate plates.) Oceans A and C are young opening oceans and their growing plates A_1, A_2, C_1 and C_2 include continents and are bounded by the trenches of ocean B which is closing. B's continental margins have been uplifted into a cordilleran mountain range in which volcanoes are still active while the margins of A and C are accumulating erosion products on a broad shelf.

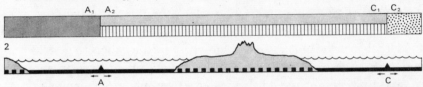

2   (50-100 million years later) Ocean B has now closed and the two cordilleras have collided to make one high mountain range, volcanically inactive and already starting to erode. Oceans A and B are still spreading from their ridges but now no trenches are available to destroy their surplus sea floor. The destruction of B's distinct plate now leads to ambiguity in defining the boundaries of plates A_2 and C_1, which appear to have combined.

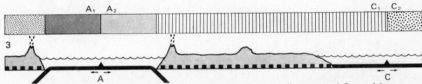

3   (Several million years later and probably after the closure of another distant ocean) Ocean A has opened trenches which are destroying the surplus sea floor produced from its own ridge and that of ocean C, clarifying the boundary between plates A_2 and C_1. Cordilleran ranges have appeared along the margins of A, and the older central range in the new supercontinent is being worn down by erosion.

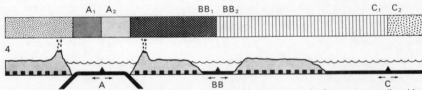

4   (100 million years later) The closing of ocean A has provided an opportunity for a new spreading ridge and ocean to appear in the supercontinent, in a slightly different position from the old suture line. New ocean BB now has two plates BB_1 and BB_2, both carrying a continent, though plate BB_2 can equally well be described as C_1.

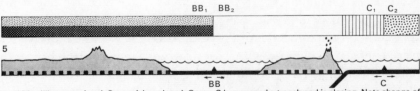

5   (100 million years later) Ocean A has closed. Ocean C has opened a trench and is closing. Note change of plate boundaries and clear ownership of continent by plate BB_2. As continents have changed direction of movement, new rocks have been added to alternate leading edges. Broken line under continents represents old basalt sea floor fragmented by eruption of continental volcanoes during earlier plate movements.

it likely to be done until Russian and Chinese geophysicists grow a little warmer to the whole idea of sea floor spreading.

It may be some comfort to all the oceanographers who are engrossed in playing with plates that the introduction of the new Dewey-Horsfield model outlined above should allow them to indulge themselves with a very flexible new set of rules for recreating the fit of continental margins. The famous Bullard, Everitt and Smith reassembly of the continents around the Atlantic and the more recent (1970) Hallam and Smith fit of the southern continents made on the same principles show how exactly the margins of continents bordering *opening* oceans can be matched, after separation of tens of millions, or even hundreds of millions of years.

In the excitement of these 'perfect' rearrangements it should not be forgotten that Bruce Heezen managed to produce a rough and ready fit for *all* the continents on the globe he painted in 1958 when, in the first happy glow of discovery, he suggested the earth might be expanding. The new model predicts a similar approximate fit for the margins of continents around *closing* oceans allowing for the disruption and distortion caused by the eruption of island arcs and cordilleran ranges. If and when detailed geological studies can make clear exactly how much new material has been added to such margins during a closing cycle then the new strips of continent could theoretically be peeled off the map to give the fit of the margins as it was in the earlier opening phase of the ocean.

Further back in time all is conjecture; the outlines of areas covered or underlain by rocks of different periods are far too indistinct to provide more than a general qualitative guide to what the map of the world may have looked like at say 1,000, 1,500 or 2,000 million years ago, though at a long shot it is possible that if we had more precise observations we should still find all the margins of the smaller, older continents fitting together much as they do today.

Exactly the same reasons that make the fit of continental margins more accurate for opening oceans than closing oceans, governs the fit of continental margins to spreading ridges. Judging by the present plate pattern the opening of trenches on both sides of an ocean is less likely to occur than on one side only. So as a closing ocean ages the position of its spreading ridge will probably become very asymmetric and in the later stages of plate consumption the ridge will vanish altogether. It seems probable that this is what we are seeing, at an intermediate stage, in the position of the East Pacific Rise today. The ten-year-old suggestion that this ridge was not formed as a truly mid-oceanic ridge may have to be discarded. Under the rules of the new model the rise could have formed in an earlier supercontinental land mass and started its life history as a normal, symmetrically placed mid-ocean ridge. Its success as a spreading ridge probably contributed to the closure of the proto-Atlantic. When that ocean had vanished and its trenches had stopped work the first Pacific trenches may have opened along the western coast of North America, bringing about the formation of the Cascade Mountains and the Sierra Nevada.

Those trenches, by consuming the north-eastern part of the Pacific plate, must have allowed the present North Atlantic to open so that North America could move far enough westwards to overrun the position both of the trenches themselves and of the spreading ridge. The trenches now consuming the western Pacific plate may not have started work until the eastern plate had almost vanished (the closure of the Urals' ocean probably complicating the sequence of movement), so giving the whole ocean basin its present lopsided appearance.

To many people this picture of the earth in action will probably seem too bizarre for belief. Even if they were prepared to accept a primeval supercontinent that broke up into drifting fragments they may want to draw the line at continents whizzing around on the globe and bumping into each other like dodgem cars at a fairground, especially dodgem cars that grow bigger every time they move. They may feel that in terms of credibility there is not much to choose between this story and the one told in the first chapter of Genesis when on the third day the waters under the heaven were gathered into one place and the dry land appeared.

Certainly to survive outside the realms of 'geopoetry' a new model like this must not only be able to make sense of observations already made: it must also maintain itself as a satisfactory framework for a whole lot of new ones still to come. Before long our model will be put to the test by the new work in ocean science as results pour off the presses in an almost continuous cascade of paper. In a matter of months new papers have challenged Hess's proposals on the composition of the deep sea floor rocks, a new model has been worked out for the preliminary stages in rift formation, and a wholly new fracture zone has been discovered. Meanwhile bit by bit thousands of square miles of ocean are coming for the first time under the drills and exploring instruments of the research ships. 'The whole north-west Pacific floor is up for grabs' as one oceanographer at Scripps put it, and when the detailed features of this huge and complex area have been identified, and the motions of what look like several ancient plates have been established, it may provide powerful evidence of continental movements in recent cycles.

Already it is clear that any acceptable history of the crust of the earth must be able to embrace all sorts of variations as well as the main theme. It must account for huge horizontal movements but also allow for large vertical displacements. Some of these are explicable when land is depressed under the thick cover laid down by ice ages and rises again slowly after the thaw with all the accompanying adjustments of sea level. Some are far less obvious, when whole areas seem to tip, or even to sink below the 'normal' surface of the sea for very long periods. According to recent findings Northern Europe is still recovering from its last burden of ice 10,000 years ago, but the British Isles are also showing a gentle tilt, while round their northern coast are large areas of foundered 'continental' land. At the same time far to the south the rocks of the Afar Triangle on the Ethiopian coast of the Red Sea, apparently typical sea bed basalts complete with magnetic lineations, rise inexplicably high and dry beside the young ocean in which they were formed.

Even when all these oddities have been accommodated to satisfy the rigours of earth physics, the new model must also stand up to the detailed findings of other kinds of work, studies of the 'provinces' of plants and animals, the detailed findings of a whole army of land geologists and last but not least of palaeomagnetists. Palaeomagnetics as we have already mentioned, is the technique by which the magnetism of the rocks is 'analysed' to reveal the disposition of the magnetic field in which, if they are volcanic, they cooled, or if they are sedimentary, they were deposited. In other words, if the orientation of the rocks themselves can be properly established, the position of the magnetic poles can be deduced.

By doing this with a lot of different rocks of different dates from the same area palaeomagnetists can draw what they call a 'polar wandering curve'. This as we saw in Chapter I means that if it were proved that the continents did not drift the magnetic poles would have wandered through all the positions plotted on the curve – and it was the unlikelihood of this that made palaeomagnetists first support continental drift. Now polar wandering curves for different continents can be compared over longer time periods. Where they match the continents were close together; where they start to deviate the continents separated. Eventually this technique, applied systematically to the 'strips' of new continental material that, if we are right, have accreted to different continental margins in turn, depending on the direction of movement, should provide one of the most powerful tests of all for our new model.

It is an alarming obstacle course for an idea but there is on the other hand some encouragement in the reflection that Oceanography has usually only ruined the reputations of people who dared to speculate too little and thought on too small a scale. She has smiled most benignly on those who backed the most daring and outrageous possibility; the adventurers who believed the fathomless depths could be fathomed with lead and line; the naturalists who believed in a whole new world of life in the deep waters; the impractical dreamers who dreamed that continents could drift: the prophets who foretold submarine mountains running in continuous lofty chains for thousands of miles; the poets who imagined the spreading sea floor.

8

Looking Ahead

FASCINATING THOUGH IT MAY BE TO DISCOVER THAT THE FAMILIAR MAP OF the world is only a kind of snapshot of the continents in movement, a record of a moment in time, obsolete by the time it is printed, it is not a matter over which anyone is likely to lose any sleep. Even if New York is half an inch farther away from Lisbon or Buenos Aires is half an inch farther away from Cape Town than they were last year it isn't going to put up the cost of the fares for quite a while yet. Even if most of California is already on its way to Alaska and the sandy beaches of France and Italy will one day be indistinguishable from the deserts of North Africa, the tourist industry is not yet threatened. Even if North America and Asia are inevitably doomed to an eyeball-to-eyeball confrontation when the Pacific Ocean has all been consumed by its trenches, there are probably a couple of hundred million years in hand before the situation becomes embarrassing. By then man and all his politics will almost certainly have vanished from the earth.

Meanwhile for most of us there is more urgent business on hand – a job to do, a family to raise, bills to pay and pressing public problems to catch our attention like the population explosion and environmental pollution. Yet, as it happens, the newly discovered facts about the ocean floor will eventually extend their influence to all such human affairs. We have claimed that oceanography is 'big science' and that it is a characteristic of big science to disrupt and modify all other sciences around it. It is impossible to move or change one large area of knowledge without making the rest of the pattern unrecognisable.

So far the changes that have occurred have probably hardly been noticed by most people. Many of them have been subtle adjustments of a broad picture of the world that is built deeply into a culture. Take for example that ancient story of Atlantis. For over 2,000 years it has been handed on from generation to generation, even among people who never heard of Plato, never learned a word of Greek, as one of the great legends of the sea.

Atlantis, the sunken continent, lost with all its people, its buildings and its 10,000 chariots and 1,200 ships, in a mysterious and dreadful catastrophe, must have haunted the imagination of millions with its overtones of loss, of waste, of guilt and decadence and of supernatural vengeance. Now, in the last couple of years, archaeologists and oceanographers between them have accumulated an impressive amount of evidence to identify Atlantis with the ancient Aegean island of Thera, the remains of which are now known as Santorini. 'In a day and a night' between 1500 and 1400 B.C. the island sustained the kind of volcanic explosion and collapse that destroyed Krakatoa in 1883, with the possibility that the Thera event was five times more powerful. There is evidence that the 'tsunamis' or seismic sea waves caused by the explosion and the earthquakes that accompanied the collapse of Thera's volcanic core when its chamber of molten magma had voided in the great eruption, were the physical agents that destroyed the great Bronze Age civilisation of Crete.

Such events, commemorated in the folklore and oral history of the Mediterranean peoples, can be considered as geologically recent incidents in the protracted drama of the closure of the Tethys Ocean and of the vulcanism and earthquakes associated with the remains of the Mediterranean trench system. Less than two hundred miles to the south lies the northern boundary of the African plate as it is overridden and consumed below the Eurasian plate margin.

There are other shifts in our consciousness of the world we inhabit even if the media are slow in picking up the details. When the inhabitants of Tristan da Cunha in the South Atlantic were driven from their island home by volcanic eruptions in 1963, oceanographers already had a clear idea that the volcano was erupting on the flanks of the Mid-Atlantic

Ridge. The echo sounders have found dozens of similar but smaller cones beneath the Atlantic waves. Nor were the oceanographers and geologists overwhelmed with surprise when the new volcanic island of Surtsey erupted from the sea just off the ridge line south of Iceland. Yet when these events are reported or even recalled they are still treated as inexplicable acts of Providence rather than new manifestations of the normal activity of an age-old system.

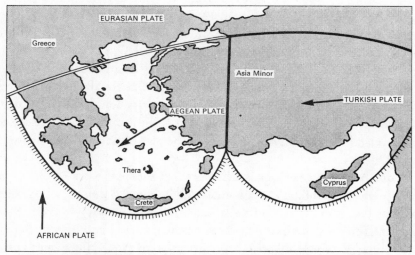

The legendary 'lost' Atlantis is now thought to have been the Aegean island of Thera (Santorini) partially destroyed in a catastrophic volcanic explosion between 1450 and 1400 B.C. The volcano, like many others in the area, is associated with the descent of the African plate below the smaller Aegean and Turkish plates, part of the process of closure of the ancient Tethys Ocean and the modern Mediterranean. Arrows show plate movements relative to the Eurasian plate. Approximate positions of the plate boundaries are shown by the double lines (spreading boundary creating lithosphere), the black line (faults) and the spiked line (consuming lithosphere).

In the long view the entire surface of the earth, continent and sea floor alike, has been produced by some kind of volcanic activity and earthquakes are an inescapable consequence of the physics of the earth and the processes that have led up to our own existence. All the same we are taking our time in accepting the meaning of our own discoveries. It has been particularly noticeable that in the reporting of the recent disastrous earthquakes that caused such tragic loss of life in Peru and Turkey and a few years earlier in Alaska, there was almost no acknowledgement of the new information about geophysical structures. Yet it is now abundantly clear that these areas are constantly vulnerable to catastrophe.

Some effects are probably obvious, particularly those that directly concern the oceans themselves. What is now known about the geology of the ocean floor is bound to influence the political decisions made about its exploitation or guardianship. If, to take an absurd example, some purely arbitrary decisions were made on mineral or territorial rights to the sea floor by simply drawing lines on a map, some countries might find themselves rich overnight in new deep sea resources, while others might find themselves lumbered with nothing but floods of basalt lavas and isolated marine volcanoes. Some might find their sea floor disappearing down trenches, however slowly, while others would find their ocean basins growing fractionally larger every year. The new information becomes more important when decisions have to be made, as they are being made now for example, in the case of Rockall, the rocky island off the coast of Scotland, as to whether a particular island or foundered land mass is a feature of the open ocean and therefore free for all or whether it is technically a part, under international law, of the country nearest to it. Rockall is, it seems, an isolated fragment of continental rock, part of the original land mass of the British Isles, a large part of which has submerged below the sea surface.

Equally obvious perhaps is the effect that the new work ought to be having on such military, economic and political decisions as whether nuclear test bombs, dangerously contaminated wastes or even plain rubbish should be put into deep ocean trenches. Or, in a different mood, whether is it worth mapping certain tracts of ocean floor with such detailed information about the magnetic signature that it could be used for navigation or for distinguishing the presence of magnetically 'foreign' bodies such as submarines.

All this is only the visible tip of the iceberg. The concept of continental drift and the elucidation of a mechanism to explain it have already caused a minor revolution in an area of thinking that supports a good deal of modern biological work – the detailed unravelling of the progressive stages of evolution.

When the Swedish ship *Albatross* sailed off on her round-the-world oceanographic voyage in 1947–48 under Professor Hans Pettersson, one of the objects of the voyage was to search for the submerged remains of 'land bridges'. Everyone thought these must once have existed to enable plants and animals to spread from one continent to another so that the same species were found in living or fossil forms in environments separated by hundreds or even thousands of miles of deep ocean. Those were the days before continental drift was respectable enough to rescue the palaeontologists from the worst of their zoogeographic problems. Now many of them are happily accepting the contributions offered by drift to explaining animal distribution, isolation, specialisation and even competition to extinction. A growing list of publications testifies to the reorganisation of the data laboriously assembled over more than 100 years.

In the conventional hierarchy of classification each class, each order and each species within the order evolves characters that enable it to fill a particular ecological niche, achieving what is called 'adaptive radiation'. At the same time another process called 'convergence' causes animals that live the same kind of life (specialising for example in eating a particular kind of plant or insect or other animal) to develop physical similarities even when they originate in widely different species or orders. Here the popular example is the group of four different species of mammals, each from a different mammal order, that specialise in eating termites; the primitive spiney anteater of Australia of the order Monotremata, the African aardvark that has the order Tubulidentata (meaning tubular-toothed) entirely to itself, the South American anteater of the order Edentata (toothless) and the African and Asian pangolin of the order Pholidota. Although one is spiney, one is scaly, one is big-eared and short-haired and one short-eared and bushy-tailed, all these animals have evolved the long questing nose essential for a creature that lives by burrowing into anthills.

In 1967 and 1969 the Finnish palaeontologist Björn Kurtén of the University of Helsinki wrote two papers in which he attempted to set out the effects on the evolution and distribution of animals of adaptive radiation and convergence combined with continental drift. He assumed that Laurasia and Gondwanaland existed as two supercontinents, one northern and one southern, throughout the greater part of the 200 million year long age of reptiles, which started in the Permian Period around 280 million years ago, and continued through the Triassic, the Jurassic and the Cretaceous period ending 65 million years ago. Although the separation of the continents had probably begun as early as the Triassic period, between 230 and 180 million years ago, Kurtén thought the rifts between them would not have become an impossible barrier to land animals until well into the Cretaceous period when the age of reptiles was almost over.

The age of mammals on the other hand, was a period when the continents were getting steadily farther away from each other, and some of the collisions that took place later had not yet happened, so there were fewer land connections between continents than there are now. If the

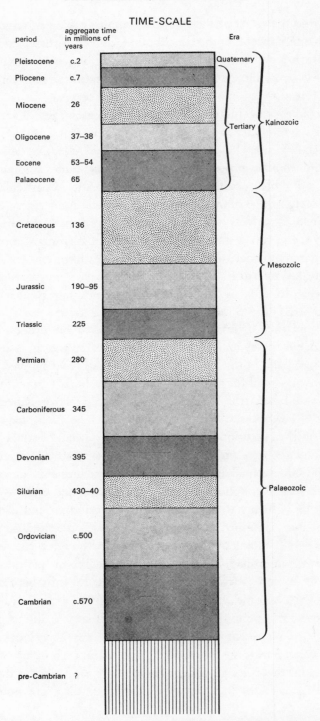

TIME-SCALE

period	aggregate time in millions of years	Era
Pleistocene	c.2	Quaternary
Pliocene	c.7	
Miocene	26	Tertiary — Kainozoic
Oligocene	37–38	
Eocene	53–54	
Palaeocene	65	
Cretaceous	136	Mesozoic
Jurassic	190–95	
Triassic	225	
Permian	280	
Carboniferous	345	
Devonian	395	Palaeozoic
Silurian	430–40	
Ordovician	c.500	
Cambrian	c.570	
pre-Cambrian	?	

The Phanerozoic time-scale, showing geological time divided into periods according to their fossil contents. The length of the earliest pre-Cambrian period, the rocks of which were until recently thought to contain no fossils, depends on the date assigned to the earth itself.

mammals began to diversify from their ancestral stock (he took the view that there was a single ancestor) in the late Cretaceous and early periods of the Tertiary (the Tertiary 'era' contains a whole group of relatively short geological periods) they would have been separated off into at least eight separate groups by the drifting continents. Their later development would therefore have been affected by a long period of isolation until the continents reached a configuration in which some new land connections were established.

Using the fossil record to try and locate the continent of origin of each mammal order Kurtén argued that the drifting continents encouraged the mammals to diversify by 'adaptive radiation' into some 30 orders in their short reign of 70 million years while the reptiles on their supercontinental land masses only achieved about 13 orders in their 200 million years. He even went as far as to wonder whether any particular province in a finite space of time *could* only 'produce and support a given amount of basic zoological variation'.

Since palaeontology is no less competitive a field than any other branch of modern science and therefore no less effective as a hot-house for academic passions it would be very surprising if Kurtén were allowed very much time for uncontested enjoyment of his new theories. Already Dr. Alan Charig of the Natural History Museum in London has stepped into the arena armed with a supply of powerful objections to Kurtén's simple correlation of the number of orders in an animal class with the number of 'faunal provinces', in this case continents, that they occupy.

Charig disagrees with Kurtén's methods of counting reptile orders, with his selection of the Cretaceous period, at almost the very end of the reptile span, as a typical time-sample for comparison, and also with his attempt to study adaptive radiation on the high level of orders instead of the lower level of family and species. He also quite rightly points out that the major part of the success of the mammals depended on such biological 'improvements' as warm blood, hair, their superior organisation and adaptability to new environments and the ability that most of them developed, by the evolution of the placenta, to nurture their young to a late stage within their bodies instead of laying eggs.

Whatever happened, and there is clearly room for plenty of argument, the rifting of large land masses such as the postulated Laurasia and Gond-wanaland into the smaller continents of today would have caused consider-able climatic changes. One way and another these could have hastened the decline of the great reptiles and especially the dinosaurs, most of which, for reasons that are totally mysterious, became extinct at the end of the Cretaceous period. Whether changes in climatic conditions and the rise of new kinds of plants, including the flowering plants, triggered the successful rise of the mammals and whether the mammals had already reached and passed a point of crisis in their development before the continents split up are all the kinds of considerations that are critical. Years of debate, of guesswork and of the discovery and presentation of new evidence stretch ahead of us if the matter is ever to be settled.

Meanwhile other traditional but rather more specialised palaeonto-logical debates are taking on a new lease of life from the revelations of geophysics. One that is already in full flower is the little matter of the marsupials. Marsupials are mammals which give birth to their young at a stage of development unsuitable for independent life so they have to be carried for some time in a pouch on their mother's body, whereas the young placental mammals are born in a more developed state.

Fossils of early forms of both groups have been found in North America dating from the middle of the Cretaceous period about 100 million years ago. It is therefore something of a mystery that the 'natural' fauna of Australia consists almost exclusively of marsupials and has no placentals. A great deal of energy has been devoted to trying to work out how marsupials, and only marsupials, got there. If it was possible to get to Australia by an easy land route it would be reasonable to expect placentals to have got there as well. If it was always difficult to get there because it meant a long sea crossing or at the very least island-hopping for a good part of the way in from Asia then either the marsupials might have got there by a lucky accident or they went there in an early form that made them particularly suitable for the trip and achieved their adaptive radiation after they arrived.

The confused evolution of the sea floor spreading theory has resulted in the confused evolution of the marsupial theory. In what sounds like a desperate attempt to crack the conundrum Dr. P. G. Martin of the University of Adelaide has recently proposed that the marsupials evolved as long ago as the Jurassic (190 million years ago) in a land mass originally situated in the central Pacific over the position of the Darwin Rise, the area of volcanic islands, atolls and guyots that led Menard a decade ago to suggest that it was the remains of an old, subsided and inactive mid-oceanic ridge. Two main fragments of Martin's land mass, split by the

Darwin Rise in the early Jurassic period, are imagined to have drifted respectively south-west to become part of New Guinea and north-east to become part of North America, each fragment carrying a convenient load of marsupials. This would indeed explain their obvious presence in Australia and Southern America, the finding of their fossils in North America dating from around 65 million years ago and in Western Europe from 50 million years ago, and finally the embarrassing absence of any of their fossils in Australia before the late Oligocene period (between 28 and 38 million years ago). Like the marsupials stranded on their isolated continental raft in the middle of the Pacific this theory, at least in this form, is unhappily isolated by the change of direction in the spreading theory away from the idea of a possible Darwin Rise at all. The idea was an early casualty of the plate theory.

So now a new proposal has been put forward by Dr. C. Barry Cox of the Zoology Department of King's College, London. He holds the alternative belief that the marsupial mammals evolved *before* the placental mammals and may in fact have given rise to them later on. The marsupials might then have been able to migrate through Africa, to Antarctica and Australia before they broke away from Gondwanaland and before the placentals reached them. (Both Australia and Antarctica have changed their positions considerably in the course of the Gondwanaland break-up: Australia has been much closer to the South Pole and Antarctica was formerly in much lower latitudes.) When Antarctica moved southwards the ice killed off all its landbased animals, so Australia, which eventually separated from Antarctica to take up its present position, became the only continent exclusively occupied by marsupials.

This theory hangs on the timing of the various significant events; the evolution of marsupials, the evolution of placentals, their respective places of origin and subsequent migration, and finally on the dating of the various episodes of continental drift.

Unless you are an Australian, or a dedicated professional palaeontologist, or at least a person responsible for the proper arrangement and labelling of fossils in a museum, you are quite likely to find arguments like this pleasantly hilarious. Seen from a certain angle it can all seem very much like the great debate on how many angels could dance on the point of a pin, that is supposed to have preoccupied the intellectual leaders of Constantinople in the fateful year of 1453 when the Turks were already beating at the gates. Yet it is just this kind of close deduction and detection that is gradually revealing the origins of man, that may unravel the mysteries of race and colour differences, that may even tell us whether our ancestral stock was peacefully vegetarian or violently cannibalistic.

The sort of evidence that is available to the palaeontologist at the moment is not by any standard all equally reliable. The evidence from the fossil record is often rather negative. Because fossils are not found does not necessarily mean that an animal did not exist in a certain place at a certain time. The only certainty from a fossil find is a limiting date by which a species is known to have existed: and to exist it must have had some evolutionary history. In rather the same way the drifted continents and the tantalising fit of their margins can only be useful within the limits provided by the few positive dates established by drilling and some palaeomagnetic studies. So much effort is going into the attempt to unravel the history of the sea floor that some geophysicists confidently predict that in about five years, by co-ordination of all the available techniques, it will be possible to produce an atlas showing the position of all the continents and oceans at convenient intervals of time over the last 200 or 300 million years, all the way back to the age of whatever turns out to be the oldest section of the present ocean floor.

When that day dawns it ought to illuminate all kinds of other studies that will help palaeontologists in what probably feels just now like a struggle to piece together a meaningful picture out of jumbled bits of two separate and defective jigsaw puzzles. Once the position of seas and continents is known it should be possible to arrive at a theoretical approximation of climatic conditions at different periods. This in turn should extend our understanding of the evolution and distribution of plants and their effects on animal populations.

On a different scale the marine palaeontologists who specialise in analysis of the micro-fossils of planktonic animals – the foraminifera and radiolaria – identifying species and subtle changes of individual biological characters with an electron microscope, will undoubtedly bring valuable new interpretations into the pictures of the zoogeographical past. Their special advantage comes from being able to find the fossil record of different species over 100,000 years in half a metre of ocean sediments so that they can detect the tiny step-by-step changes, each one within the limits of individual variation, that lead either to extinction or to successful adaptation to environmental changes. The fossil material in ocean sediments was deposited with unfailing regularity as the season of productivity came and went with every passing year. Their time-span and continuity is unparalleled in any part of the fossil record on land, where preservation of specimens depends on the chance of such irregular disasters as a land-slip, the sudden effects of a flash flood that swept animals away into a pile of mud and debris, a fatal blizzard that caught mammoths at their pasture, a treacherous bog or quicksand or asphalt pit that engulfed the unwary sabretooth or overweight dinosaur or preserved a woolly rhinoceros in a bath of sand and oil.

As long ago as 1950 this kind of study established that a change of water temperature of as little as 2° or 3°C is enough to account for the change in an entire marine planktonic species from a left-hand shell whorl to a right-hand shell whorl. Studies of populations and the statistical techniques that can interpret them are already being applied to try and detect the effects of the opening or closing of particular oceans. In the course of its opening an ocean would become deeper and colder, probably linking up with deep cold water from other basins and the succession of its fossils would reflect each stage in its history. The work is slow and meticulous but it is gradually accumulating results. Eventually enough will be known of the biological composition of fossil plankton shells and their relationship to various parameters in their marine environment to lead to a clear understanding of how certain species become extinct and therefore, by deduction, how drastic were the changes sustained by the environment, both at sea and on land.

It would be a mistake to infer from all this that the movement of ideas from the oceanographers (or more particularly from the proponents of sea floor spreading and plate tectonics), to the palaeontologists and land geologists is a one-way traffic. From its earliest beginnings the very notion of continental drift leaned heavily on the geological and palaeontological evidence, long before magnetics and deep drilling came to its assistance. Even now, with 10 years of consolidating effort backing up the spreading theory, it was the discovery in Antarctica in 1969 by a team from Ohio State University of a fossil skull of Lystrosaurus, a reptile which has also been found in the Triassic levels in Asia and Africa, that for many people put the seal of authority on the story of drift. With the exception of the 'Piltdown Man' with its ambiguous overtones of fraud, practical joking and academic vengeance, a fossil is incontrovertible evidence.

The relationship between the geophysicists and other scientists should perhaps be seen today as a complex system in which each incoming piece of new information or interpretation modifies the programme of investigation of all the individuals involved. With luck such a system, in which the special insights and demands of many different disciplines are even roughly co-ordinated, may soon make discoveries that will totally change our understanding of the evolution not only of ourselves but of our whole physical environment, and may even, in passing, help us to preserve the gigantic, subtle ecological network in which we exist.

It would be foolish to predict what kind of results might be achieved in even a few years by such a powerful new interdisciplinary research tool. In the short term it should be safe to forecast that when the oceanographers have presented their final report on continental movements over the period of time represented by the age of the present ocean floors, now roughly dated at the start of the Mesozoic era, the frontier of enquiry across all the associated sciences will move 300 or 200 million years further back into the past, and the searchlights will be turned on still darker ages of the earth's history. If the present pattern of tectonic plates provides the key to the structure and growth of the continents it might even give palaeontologists some unforeseen new clues in interpreting the pre-Mesozoic history of flora and fauna, perhaps in tackling evolutionary problems of bony fish or amphibian distribution before the rise of the reptiles!

It is not even necessary now to apologise for a digression into speculative fantasy by following the trail broken by sea floor spreading back to the very beginnings of life on this planet and perhaps even further, to other planets and worlds in the inaccessible depths of space. If we consider seriously the possibility that the continents as we know them are the 'secondary' products of plate action, then we also accept the possibility that there was a stage in the history of the earth when no continents existed because the movement of the plates had not yet started to make them.

This state of affairs means that for some so far unknown period of time the continental rocks, that we take for granted as available materials for the chemical processes of living things, had not yet appeared, even if the sea itself was already in existence. In the observable present the volcanic rocks produced at spreading ridges and even in submarine eruptions on their flanks are dense and basic in character while the volcanic rocks associated with descending plate movements are on the whole less dense and more acidic. As a general distinction the continental andesites and the granites are not found on the sea floor or on marine volcanic islands like Hawaii;

they belong to the island arcs and to the great continental cordilleran ranges.

Questions then arise about the extent to which the organisation of pre-living chemical structures are dependent on materials that only continental rocks could supply. Chemists involved in unravelling the complexities of 'chemical evolution' that led up to the production and reproduction of the successful living cell are given to referring in a matter-of-fact way to the geological conditions of the primitive earth and even to a 'geological origin' of life. This kind of assumption dates back to the work that followed the proposal of Harold Urey in 1952 that the early atmosphere probably consisted of a mixture of methane, ammonia, hydrogen and water vapour. Stanley L. Miller, his associate at the University of Chicago, put this gaseous mixture into a flask and supplied it with energy from simulated 'lightning flashes' of around 60,000 volts. After a week of this treatment the gases had produced about four of the 20 amino-acids that are the building bricks of protein in living cells. Other researchers in other laboratories were soon hard at work trying to improve on these findings, by modifying the gas mixture, by changing the energy sources, finally by trying out various additional substances as catalysts to speed up the chemical reactions, since after all the atmosphere has a close relationship with the surface of the earth.

Sidney Fox, now the Director of the Institute of Molecular Evolution at the University of Miami, conducted a series of experiments along similar lines but using thermal heating instead of an electric discharge as the energy source. To assist the chemistry he arranged that various catalysts were tried in turn by using them as packings in a hot furnace. He tried both alumina and silica as catalysts and found that silica, especially in the form of quartz sand from a Florida beach, produced an almost complete array of the amino-acids required by living organisms.

It was assumed that the materials used as catalysts, since they were all readily available on the surface of the earth today (in fact they figure largely in its composition), were equally available on the primitive earth. The point is that they may not have been. Just at the moment no-one, certainly no geologist, would pretend to know what the primitive crust of the earth *did* contain – for that we must await the pronouncements of astronomers and cosmologists – but quartz, from which Fox procured his quartz sand, just *might* not have been lying about on beaches and mountain tops if the continents had not already formed. The rocks of wholly 'oceanic' volcanoes would not necessarily contain it nor would the rocks of the sea floor if those features were similar in composition to their counterparts today.

It is true that some rocks of a granitic, acidic affinity have now been dredged up from ocean ridges and fault zones, but though their presence may be the result of the fractionation of cooling magma they seem to be fairly uncommon. They certainly could not be depended on to provide the same chemical effects as continental quartz in the (perhaps unlikely) event that this particular catalyst, in generously available quantities, should turn out to be essential to the biological recipe.

This argument also applies to the 'silicic' cherts found in the ocean associated with the sediment layers – the same flinty material that we have met obstructing the drill bits of the explorers – and which make up some one per cent of the volume of volcanic rocks in Iceland, generally regarded as a wholly oceanic outcrop of the Mid-Atlantic Ridge. The fashionable description of cherts as the result of chemical action on the siliceous shells of planktonic animals does not explain the ancient examples found among some of the oldest continental rocks in the world, dated at more than 3,000 million years old, long before any true living organism existed, let alone anything as advanced as the plankton. These cherts at any rate must have been formed out of some other kind of sediment. Whatever their origin the cherts of 'modern' oceans are formed in very deep water and it would be most unwise to presume that the ancient examples were not also laid down at the bottom of whatever ocean existed at the time. Apart from occasional outcrops it is only when areas of sea floor are cut off by continental structures that the cherts are likely to have been uplifted into the regions of shallow water or air where it is thought the essential chemistry of life began.

Establishing that something could happen does not prove that it did happen and 'molecular evolution' has long ago moved on past its early pre-occupation with the possible mechanism of the formation of molecular chains into areas of fascinating complexity but just conceivably it may still have to pause and consider whether, if continents are not a primeval feature of the earth's crust, the synthesis of amino-acids might have had to wait for their arrival.

Certainly these dangling links and shadowy possibilities are at the moment only ideas. At best they are on the distant horizon of research; but the advances of space technology have already given an enormous impetus to chemical evolution, pre-biotic chemistry, molecular palaeontology and similar studies. Sending men to the moon or to the planets or introducing to the ecology of the earth materials brought back from these alien environments has always been seen as a biological hazard and men and funds have been available to get its investigation moving at a fairly fast pace. So whether or not we are yet within sight of finding out how the chemistry of a living cell got itself organised, we are very likely to witness some astonishing discoveries as we piece together the conditions that made life possible. It would not after all be too surprising if we found that its very development on this planet was even more closely connected with the development of the continents than the implications of later drifting suggested.

Further back still who dares to venture? Space scientists and geophysicists do. Already a sizeable collection of papers have appeared discussing the possibilities of plate action on the moon. The debate centres on the mysterious 'mascons' (contracted from mass concentrations) discovered when the orbits of the early moon probes were distorted by unexpected gravitational effects. These concentrations, generally associated with the *maria* or 'seas' of the moon, were consistent with the presence of large dense bodies beneath its surface and would need to be explained if the moon were nothing more than a ball of dust and cosmic rubble. Such lumps might of course be the remains of huge meteorites or similar objects embedded deep below their impact craters but if so the traces of their arrival have mysteriously vanished.

At the time of writing hundreds, indeed thousands, of scientists are still grappling with the enormous and possibly insoluble problem of deducing the entire structure and history of the moon from a few bucketfuls of surface rocks and dust brought back by the Apollo 11, 12 and 14 astronauts. For the mysterious mascons to be nothing more interesting than sheets of dense lava demands some kind of hot moon. For them to be plates demands a degree of organisation in the moon's interior for which there is so far no evidence.

It seems, however, that the moon rocks are igneous and that the lunar surface at least has at some time in its history been hot. It seems also that certain of its rocks carry the imprint of a 'fossil' magnetic field far stronger than that present on the moon today. So the discussion of possible moon plates is by no means closed. Even if they do not exist and the moon turns out to offer no more than a miniature history of the earliest pre-plate stage of the earth, we may well find plate development on other planets in the solar system.

Putting all these speculations and new concepts into a bag together and shaking them up may give us an even more remarkable pattern of development of our world and ourselves from what may have been their original particles of interstellar dust. Looked at from one angle evolution can look like the inevitable adaptation of chemistry to certain limiting conditions; from another angle it can look like an almost purposive line of progress.

Put another way, if we gaze backwards from today at the possible beginnings of the planet, life seems to have made its way forward through a 'window' in a series of otherwise opaque and impenetrable barriers. The more we discover from the earth sciences, from the life sciences, from physics and from mathematics the longer we have to make the list of apparently accidental conditions that permit life on earth to survive, even to get started. Although other kinds of life may be possible, perhaps based on elements other than the carbon of organic chemistry, we know that our kind of life could not exist unless the size and mass of the planet was adequate to retain an atmosphere and certain essential gases in it as well. We know that the behaviour of this atmosphere must filter out certain harmful radiations like ultra-violet light that would kill us if it reached the surface in its full strength. We know that the temperatures on earth must allow water to exist in its liquid and gaseous states and that unless all these conditions are met, then life might not occur. Now perhaps we may have to add to them motions and temperatures of the planetary interior that favour the formation and development of plates and continents.

Even at their least speculative the arguments of space scientists sound like flights of fancy so perhaps we should bring the subject of plate tectonics back to earth where at the moment it belongs and where immediate work is likely to start on the practical level of trying to turn the new geological concepts into money. It may be years before academic geologists rearrange all their existing data within the new geophysical framework but it may not be long before the structural ideas are put to the test.

Nobody need be surprised that the first attempt at a large-scale geological treasure map has come from 'Tuzo Wilson country', Erindale College at the University of Toronto. There Dr. Arthur Raymond Crawford has recently organised his own observations from India, Ceylon and Australia within the framework of a reassembly of fragments of the

Gondwanaland continental complex. Each of these countries contains
known sources of commercially valuable minerals and if they were all at
one time part of the same land mass there are very good grounds for
expecting some of these minerals to turn up in the matching rock structures that cross the new continental boundaries. The fit of the continents
becomes absolutely crucial if the geologist is to track the mineral bearing
formation from one continent to another. Paradoxically it is only when
matching formations have been located that the continental fit can be
properly established.

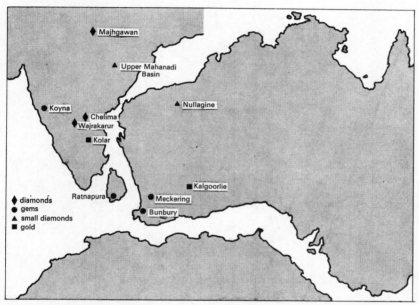

Sketch map of southern continents based on assembly by Raymond Crawford, showing possible
relationship of areas containing valuable resources in India, Australia and Antarctica.

Dr. Crawford's own work led him to draw up a tentative match that
puts Western Australia alongside Southern India and Ceylon and makes the
Kimberly Block in Northern Australia adjacent to Burma. The southern
coast of Australia then fits into East Antarctica. In such a fit there is a
special significance in the fact that the gold mineralisation of Mysore is the
same age as that of Kalgoorlie. In the same way the diamonds at Majh-
gawan in North India and in the Ananatapur and Kurnool districts of
South India suggest that Nullagine, the only place in North-western
Australia where diamonds have been found, ought not to be the country's
only fruitful site. With this arrangement there is, however, the possibility
that some Indian diamonds were actually formed in what is now Australia

and carried off into India in the course of erosion. The original position of Ceylon also becomes very significant for Australia if there is any likelihood of matching the island's precious gems and graphite deposits with Australian rocks.

Published almost simultaneously with Dr. Crawford's paper was a new computer fit of the southern continents by Dr. A. Gilbert Smith of Cambridge, one of the co-authors of the earlier Atlantic reassembly, and Dr. A. Hallam of Oxford. Their Gondwanaland reassembly is significantly different from Dr. Crawford's because it is now Antarctica that

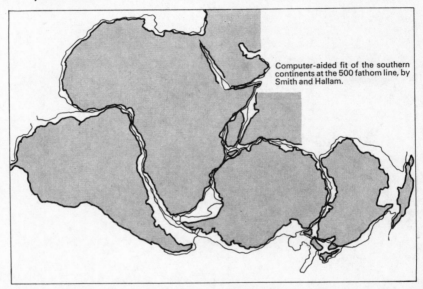

Computer-aided fit of the southern continents at the 500 fathom line, by Smith and Hallam.

lies closest to the rich diamond mines of South India, pushing Western Australia farther north towards Burma. Madagascar fits snugly into a hole between South-west India and Africa and Ceylon was popped into the remaining gap not by the computer but by eye.

Although some people may already be trying to find out how to collect the fortunes in gem stones and minerals that may be under the Antarctic ice it is probably premature to stake one's shirt on the guesses of a computer. Possibly neither of the recent Gondwanaland arrangements is right and the final answer may have to wait for the combination of detailed sea floor spreading history and palaeomagnetic evidence of the former position of the continental rocks.

Whichever geologist gets the map right and follows his judgment with some practical exploration may well retire from the circle of his academic friends to open a bank, for even without thawing the polar ice cap it will be sheer bad luck if straightforward reassemblies of this kind do not lead to some spectacular strikes of new wealth across the margins of all the southern continents.

This kind of mineral hunting is not all that the new kind of geology can offer. It can perhaps promise some minerals from the sea itself. In the last few years a considerable amount of excitement has been generated by the discovery of layers of mineral-rich sediments as much as 300 feet thick in deep valleys along what is believed to be the embryonic mid-ocean ridge of the Red Sea. The sediments contain gold, silver, iron, manganese, copper, zinc and lead and they lie under 'pools' of exceptionally hot, salty water. Under just one 'pool' the gold and silver in the first 30 feet of sediment alone are estimated to be worth 2,000 million dollars.

These extraordinary pools of water and the rich sediments below have been named 'brine sinks' and later on we shall be looking in some detail at the theories of their formation in connection with the possibilities of ocean mining. The point is that a mechanism seems to exist by which favourable parts of the new ocean ridges, where the structure and configuration of the volcanic rocks is suitable, can concentrate minerals into large and valuable bodies of ore.

The promise of such concentrated wealth is distinctly tantalising to mining engineers who have not at the moment any means of extracting it. There is a possibility, however, that the discovery of the brine sinks could take some of the chanciness out of the mineral prospecting on land. The old mining proverb that 'minerals are where you find them' may lose some of its force if the new geology based on oceanography provides a broadly coherent system to guide prospectors in their search. Minerals may turn out to be exactly where we should expect them.

The process of sea floor spreading will eventually carry these ore bodies away from the ridge crest, probably split into two sections by the next intrusion of molten basalt in the rift. Hundreds of millions of years later, if we are right in thinking that the movement of ancient plates has built up continents by collisions, the opening ocean will become a closing ocean and the ore body will find itself at a continental margin. It may be that bodies of ore rich enough to have ransomed all the kings in history have simply disappeared into trenches and returned to the mantle. It may be that they have melted and found their way, as veins of metal, into the continental rocks above the descending plate margins. It may be that ore concentrations of this sort are so much a feature of the ocean floor that wherever oceanic rocks can be found on land they will be likely to contain mineral wealth. The places to find them, if they are there at all, will be along the lines of suture or junction where continents have collided with other continents, or with island arcs, to mark the site of a totally consumed ocean. There we should find the last remnant of the ocean floor squeezed between the continental blocks in the characteristic sequences of rock types which geologists call the 'ophiolite suites'. Not all geologists would agree that the ophiolites *are* the remnants of oceans, although their presence on land, so unlike ordinary continental rocks, has been the subject of a great deal of argument and guessing.

Even if the 'brine sink' kind of ore concentration is comparatively rare, the ophiolites found in mountain belts are frequently rich in base metals and it may simply be that these are the result of fractionation of the layers of basic intrusive complexes among or below the surface basalts of the ocean floor. Geologists are already wondering which way to interpret the unexpected deposit of copper found by the *Glomar Challenger* when she was drilling in the deep ocean rocks off Cape Hatteras. In the long run, if and when economic conditions encourage mining technology to develop to such an extent, we shall probably be able to retrieve large mineral deposits straight from the ocean bed where they will have been identified either by drilling or some less time-consuming method of prospecting. In the meantime the traces of the ancient sea floor among the continental mountains should be a worthwhile guide for land prospectors.

Even before any significant changes are made in exploration for metals there are likely to be further developments in the search for oil as a result of the findings of oceanography. Although the oil industry is employing an army of chemists there is still room for some doubt about the origin and history of the oil itself. The general opinion is that it forms from the remains of marine plants and animals, many of them organisms of the plankton, which rain down to the bottom in a more or less continuous stream of debris from the biologically productive areas of the surface waters. The hydrocarbons produced by this decay collect, below layers of impervious rock, in layers of porous limestone and sandstone, in folds and traps where the oil floats above a sealing level of water.

When oil-bearing rocks are examined it seems that the oil found in formations laid down in the geologically recent past, up to 30 million years ago, are all in the tropics. Oilfields in higher latitudes, either north or south, are all in older rocks, often containing signs of warm water corals and, like the scars of glaciers in the hot southern continents, they signal that unless the climate of the world has undergone a drastic change the continents have drifted, carrying oil laid down in tropical waters to its present position. If it is found that oil *can* only form in the tropics, the oil prospector might as well ignore the sedimentary rocks formed in high latitudes, even if biologically productive areas of water do occur there, and concentrate on those rocks that can be tracked back to the tropics by continental reassembly or palaeomagnetism. Rocks or even sediments laid down in high latitudes will show magnetic dip while those laid down near the Equator will not.

OIL TRAPS

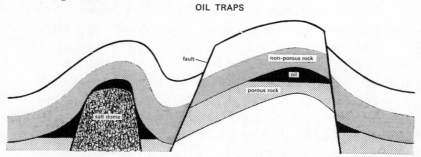

Diagram of oil traps formed by movement of sedimentary rocks and the uplifting of salt domes.

The new geology can take the oil prospector even further on his way to success. It was noticed long ago that the sedimentary oil-bearing rocks are often associated with large lumpy deposits of salt, called salt domes or diapirs, steep-sided structures that distort the sediment layers above them, by making the hollows and cavities that trap the oil, and the domes are often capped with rock containing sulphur. Recent seismic surveys above the Gulf of Mexico and the area around the Cape Verde Islands on the other side of the Atlantic have revealed salt domes rising up under 3,500 metres of water and five kilometres of sediment. Experimental drilling was done over them by the Joides Deep Sea Drilling Project investigators. One dome in the Gulf of Mexico showed traces of oil in the cap-rock and the drillers wisely decided to withdraw before they opened what might be a large oil reservoir in deep water with no chance of either winning the oil or stopping the leak.

Similar domes that could also be salt have been located by the deep seismic refraction surveys in other parts of the Atlantic, for instance off northern Morocco and southern Portugal, under water almost 4,000 metres deep and four kilometres of sediments and there may be others in the Bay of Biscay and the western Mediterranean. The average diameter of the Moroccan and Portuguese domes is 12 kilometres and they have nearly vertical walls but domes are often smaller and their steepness varies.

In a paper published in 1970 three investigators, Guy Pautot, Jean-Marie Auzende and Xavier le Pichon, working from the *Centre Océano-logique de Bretagne* in Brest, France, suggested that the salt domes were the result of thick salt deposits laid down in the initial stages of the rifting of the Atlantic, the Gulf of Mexico and the Mediterranean basins. The salt was laid down in shallow water on oceanic crustal rock, possibly even in the central part of the rift which in its earliest stages is thought to be a broad elevation. This crust later subsided to become deep sea floor as the rifting spread and the oceans became wider and the circulation pattern changed. Now the salt is preserved at the continental margins together with the sediments and the oil that formed when the oceans were narrow, warm, shallow and highly productive seas.

Tracking down the salt domes probably means that it might be possible to find new deep water deposits of oil at other rifted continental margins and not only on those which have separated in the present cycle of continental drift. Once again if we are right in our proposal that continental separation and collision is a continuous process, there will be yet another incentive to identify the site of earlier collision, the pattern of the trench systems that destroyed the closing ocean basin, and the fate of the sediments laid down in the earlier rifting stage of its life span.

If politicians have a habit of falling into a deep sleep at the mention of fascinating new ideas about dinosaurs they spring up again, wide-awake when the talk moves to exploitable mineral resources. Even if oceanography was a simple business that did not require expensive research ships and men and development funds and data processing equipment, the scientists would hardly have been able to keep to themselves the news that they had found money at the bottom of the sea.

9

The U.S. Conjuring Trick

AT ABOUT THE TIME VINE AND MATTHEWS WERE HAVING THE INSIGHT which proved so momentous in their own field, another thought, which was to prove no less momentous for oceanography, was forming in the mind of Mr. Robert MacNamara, then U.S. Secretary of Defence. Mr. MacNamara was pondering on the problem of TFX, a new aircraft for the U.S. armed forces. He had formed the view that it ought to be possible to build an aircraft which would be, as the trade jargon puts it, all singing, all dancing. It would fly high and supersonic with bombs, missiles, or reconnaissance packs, it would fly low, hug the ground and get under the radar defences of sophisticated enemies like Russia, it would swing its wings forwards to fly slowly and economically on fleet patrol, it would also land on carriers, it would help the troops with ground support fire and it would have a huge ferry range enabling it quickly to put its sophisticated finger into any pie no matter how far away round the world. The basic aircraft would appear in all sorts of marks and would be ordered in huge quantities.

The U.S. aerospace companies were sceptical that it could be done but they were appalled at the thought that most of them, necessarily unsuccessful in the competition, would be left to a life of profitless sub-contracting. If there really were going to be very few kinds of aircraft in the future then they would have to merge or diversify. In the end their worst forebodings were not fulfilled and by 1970 once again the Defence Department had returned to the doctrine of horses for courses. The main reason was that the TFX contract, given to General Dynamics, produced the unhappy F111. This aircraft, after a long history of weight problems and difficulties with engine intake geometry was abandoned in its Navy version. Most other versions proved to be adequate enough in spite of the wings occasionally falling off but the savings from having commonality of parts were less dramatic than Mr. MacNamara had believed would be achieved.

Nevertheless between 1963 and 1967 America's aerospace companies other than Douglas and Boeing (which had more than enough work with passenger aircraft and space projects) turned their attention to oceanography. Companies like Lockheed, North American, and Grumman looked to the oceans as a suitable place in which their systems capabilities could be employed. Ocean engineering on the face of it required aerospace talents, there was the problem of remote control, of hostile environments especially as far as corrosion resistance was concerned, there was a need for endurance, for lightness and strength and for reliability in manned vehicles, it was an arena for advanced technology. Furthermore it was known to be regarded with fiscal sympathy by the U.S. government.

These companies had another crucial talent. They commonly supplied the U.S. government with equipment and they were adept at persuading the defence establishment that it should support advanced technology. This was after all the only way of being quite sure that America stayed ahead of the Russians. If things were technologically possible they were likely to be done, and it was almost always possible to discover some defence significance. These companies were at the heart of the military-industrial complex and they set their powerful public relations machinery to work in this new field. As soon as the outlines of the new bandwagon could be even faintly discerned other major companies like Westinghouse swung into action as well; also on the scene, though in a cautious mood, was the Convair division of General Dynamics and even more cautious, Westinghouse's great rival, the General Electric company. Naturally the biggest U.S. manufacturer of submarines, Electric Boat, another division of General Dynamics, had to get into the picture as well.

It was not at all obvious what the aerospace and advanced technology giants would do. One cynical observer of the U.S. industrial scene said of Westinghouse 'They put 40 Ph.D.s in a room full of magazines and call it ocean capability.' In fact the teams of oceanographers and marine engineers set about acquainting themselves with the possibilities offered by the 'last frontier' or 'inner space' and they began to create projects suitable for financing by the U.S. government.

There were many possibilities. Man must learn to live under the sea they said, here he is exploring space and aiming for the moon when he cannot colonise his own continental shelf. One publicity stunt pointed up the moral. Ex-astronaut Cmdr. Scott Carpenter became an 'aquanaut' in the Sealab 11 experiment and talked to colleagues in orbit from 200 feet down on the sea bed off San Diego. The aim of the Sealab experiments was to explore the problems of living for long periods in high pressure environments underwater. It was financed by the U.S. Navy and shone in the reflected glory of space exploration.

Two disasters helped to speed things along. One was the loss of a hydrogen bomb off Palomares in 1966 and the other was the mysterious destruction of the submarine *Thresher* in April 1963. Parts of the *Thresher* were found by *Trieste II*, a bathyscaphe originally built by the Swiss Piccards. These events reminded the United States that it was very difficult indeed to find things lost on the ocean floor and that what capability the Americans had got was owed, in part, to foreign technology. Westinghouse Underseas Division for example not only had to lease Cousteau's *SP 3,000* Diving Saucer to do its own learning but commissioned the French to build an advanced 4,000 foot depth submersible to its own design. Even Grumman's recent successful exploration of the Gulf Stream with the *Ben Franklin* owes much to the Swiss.

There seemed no end to the projects that could be undertaken. At the Marine Technology Society in 1964 Dr. Athelstan Spilhaus pointed out the similarity between the ocean frontiers of today and the land frontiers of the last century. First came exploration, then occupation, then man could conquer the oceans. If exploration was the first word then great vistas opened up.

Mr. MacNamara's views on the procurement of military aircraft had therefore led with inexorable logic to teams of very capable men all over the United States, publicists, managers, engineers and scientists, sitting down to learn about oceanography. They talked, as they were accustomed to, about national goals, about capability, about development of technology, in this case deep ocean technology, and the audience was Congress and the American government.

Congressmen in particular listened attentively, especially those from states bordering on the sea or the Great Lakes. Of the 435 U.S. representatives 357 come from such states.

As things turned out, because of the Vietnam war, strong pressure to attend to social problems in the cities and the absence of international competition, the U.S. government did not fuel the ocean bandwagon with anything like the amount of money the ocean technologists wanted. Nevertheless by 1970 the critical task of education had been accomplished, the United States as a whole had become aware of the ocean frontier.

The interest of aerospace companies provided advertising revenue for new magazines specialising in oceanography. This was particularly important since specialist magazines running on a closed circulation basis are probably the most important communications channel in the United States. Television is a wasteland, radio is worse and newspapers are parochial and lack nationwide circulation. The specialist magazines, however, create worlds of their own, the doings of great companies are chronicled, the half-formed policy attitudes of indiscreet executives are written up as cliff-hangers, competitions for contracts become whodunnits and gossip is faithfully itemised. Good feature-writing advanced the general state of the art and above all else the magazines created the illusion of a new industry with its own identity. *Undersea Technology* for example says that it is aimed at 'management, educational institutions, scientific, engineering and technical personnel in industry, government, and non-profit-making research organisations concerned with . . .' then follows a list, a very long list, which is itself thought-provoking, of the items it may cover. They range from 'research, development, application, procurement, specifications, operation and maintenance of . . . underwater vehicles, floats and structures, propulsion systems, weapons, electrical power systems, checkout and test equipment . . .' and so on and on.

Is it a conjuring trick? Not really, it is no more a conjuring trick than say a banking system where paper banknotes depend on the creation of confidence for their utility. Ocean industry magazines help to create the analogous feeling of confidence, the feeling that 40 Ph.D.s in a suite of offices really do represent an entry ticket to a whole new field of profitable activity. The profit had to come from government contracts first but the corporation public relations men had already set about that problem in their own devious way.

In 1964 the U.S. National Academy of Sciences published a report on the economic benefits to be gained from ocean research. The key notion was the cost benefit ratio and it was claimed that over a period of 20 years investment in selected areas of oceanography would produce a return three times greater than the same sum invested at 10 per cent. In fisheries for example the ratio of benefit to cost was 5·8; it was suggested that over 15 years a modest 50 million dollars a year invested in fisheries research would raise domestic fish production by 380 million dollars a year. A consideration of undersea mining also produced a remarkable benefit cost ratio, 3·7. In 10 years the report said 125 million dollars of extra production could be achieved for 11 million dollars a year in research development. As if to prove its feet were firmly on the ground the report prophesied a gloomy benefit cost ratio of unity for America's decrepit shipbuilding industry. Long-range weather forecasting and least time or least cost track navigation however proved to be economically attractive. If long-range weather forecasting could be achieved for a mere 25 million dollars a year in research, 2,000 million dollars might be saved.

Even in 1963 the U.S. government had raised its ocean science budget by 52 million dollars, a sum already greater than total British expenditure. In absolute terms U.S. expenditure was 40 times the British expenditure. American industry could be forgiven for feeling bullish about the prospects and the American government was apparently a willing patron.

In September 1965 the magazine *Missiles and Rockets* produced a special edition on oceanography which appeared to mark the turning point in the rise of the new big science. Aerospace employment was levelling off, the magazine said, although the Vietnam war continued to absorb production effort. The rest of the edition was devoted to a rosy account of the expanding horizons provided by a conquest of the ocean. There was a great shortage of manned submersiles, instruments were not good enough, work was held up for unforeseen design problems to be solved. By 1975 'underseas efforts will account for 10 per cent of total sales' said Westinghouse in what proved to be a notoriously over-optimistic forecast. But the Americans broadly agreed with the Congressional witness who declared that the 1958 Geneva agreement, giving nations the right to explore their own continental shelves, had been, since it came into force in 1964, 'the starting gun of an international race to develop deep ocean technology.'

Other more diversified companies included Westinghouse, General Motors and Reynolds Metal. All these companies were spending over a

million dollars a year of their own money in improving their own ability
to compete for the supposed riches soon to come from ocean industry.

Meanwhile America's youth was listening with interest to the politicians' speeches and to the loud calls for greater participation. High school
children read the advertisements, they went scuba diving and they kept
themselves up to date with the *Scientific American* and other periodicals
which retailed the amazing progress in earth science which the marine
geologists were making month by month. From 1961 to 1968 college
enrollees for ocean science rose from 262 to 2,647 per annum.

Prof. Menard of Scripps cast a statistical eye over the scene. He concluded that the number of people at work in the field of palaeomagnetism
and marine geology alone was doubling every four years or so and that the
growth of these two sciences was the fastest in view in the sixties. If you
wanted to find a stagnant science it was, amazingly, physics.

What did the big companies think they would get out of the ocean?
They none of them really knew but they clung to the key word, exploration. An exploration vehicle seemed to most of them to be a suitable thing
to begin with, a proper underwater device that could get a man right down
there and let him get to grips with things. A research submarine had a
comforting similarity to a space capsule. Above all it was a master sales
ploy. At Westinghouse where the *Deep Star* submarines are made they are
quite clear about the psychology of it all. 'One dive equals one convert
equals one customer. Lead time? Three years, yes sir, going down there
really makes people think.'

The new wave of ocean technologists looked with pity on the old style
oceanographers from universities and institutes who ran their research, if
not on shoestrings, certainly on nothing more expensive than thin wire
lengths reeled over a ship's side. With what always seemed to be primitive
equipment these men had made astonishing progress. How much more
progress would be made with a proper plan for national action?

Public relations pressure on Congress produced 'the Marine Resources
and Engineering Development Act of 1966'. Public Law 89-454. In this
new Act there was a call for a commission to be appointed, named after
Julius A. Stratton who was chairman of the Ford Foundation. It had a
staff of 35 and drew 'material assistance' from 1,000 people. For two years
it ploughed through what there was to know. The full commission itself
met monthly for between two and four days at a time. Special acknowledgment was made to the National Security Industrial Association's

Ocean Science and Technology Committee for its series of reports on ocean user industries and to the Oceanic Foundation which organised a special support group. These and other contributions 'all provided as a public service at little or no cost to the government, were of enormous benefit to the commission.'

The Commission's report came out in 1969 and is slightly larger than this book. It proposed among other things, the creation of the National Oceanic and Atmospheric Agency (N.O.A.A.) – 'To mobilise and impart energy to the total undertaking . . .' It recognised that America's marine interests are 'vast, complex, composed of many critical elements and not susceptible to similarity of treatment'. Overall it proposed that expenditure be doubled by 1980 to 1,800 million dollars per annum.

It said other things the big companies wanted to hear. 'The Commission recommends that a framework of policies and laws be established that will allow predictability and, therefore, increased confidence and investment activity by industry.' In discussing a proposed Continental Shelf Laboratories National Project it asserted that the transfer of newly won technology to industry will be critical in advancing 'the nation's capability at sea'. It stressed the need to award survey and development contracts to industry to foster company growth and affirmed that continuing contractual relationships with large firms was the safest way of disseminating secret information. Any patents that the proposed N.O.A.A. technology programmes turned up were to be administered so as to be attractive to industrial contractors.

Nowhere in the Stratton report is there a mention of Westinghouse, Lockheed and the rest, the proprieties are strictly observed. Only Federal and State agencies are mentioned by name. The whole exercise was a triumph in public relations and corporation pressure politics.

That does not mean that the whole exercise of interesting the Federal government in oceanography was what might be called a con-job. There was probably very little intentional deception and the commission was sufficiently high-powered to see through any that was attempted. The underlying ethos was a sort of reverse application of the famous dictum 'What's good for General Motors is good for the United States.' In this case by careful selection of topics, by selective application of public relations research effort, a new arena of activity was identified and shown to be good for the United States. Therefore 'What's good for the United States is good for Lockheed, Westinghouse, and General Dynamics', because these companies already believed in oceanography and felt that they could contribute to their profits and to the industrial muscle of the United States by arranging for ocean industry to expand.

Whether it is the priority to which America should be paying a billion dollars worth of attention is another question. If it all sounds a bit of a fix to British ears this is because politics in Britain do not work, or did not work quite like that, and because people in Britain are suspicious of big science tied up with big industry. Many would see that style of establishing national priorities as inherently dangerous because it ignores issues like the plight of the poor which may turn out to be politically much more important. In fact when the Stratton report came out America was so preoccupied with Vietnam and law and order that its impact was muted.

That then was one strand in the complex weave of American policy. Another strand of comparable importance was the U.S. Navy.

In terms of economics the support of oceanography by the U.S. Navy has always overshadowed everything else, because the Navy paid for 75 per cent of all the oceanography that went on in America. As we saw earlier, the U.S. Navy had in its possession as long ago as 1962 magnetic anomaly patterns on the Atlantic Ridge. If the significance of the reversing fields had been appreciated it would have been the U.S. Navy which gave the great boost to marine geology and made continental drift respectable, not a couple of Cambridge geologists.

The Navy's activities are inspired by defence, but using the magic word capability, it can do virtually anything it likes from ocean floor topography to training dolphins. How far it strays from the straight and narrow path of anti-submarine warfare depends on the budget strings. The U.S. Navy, like the Royal Navy, learned the lesson of not neglecting fundamental research during World War II when it had to draw heavily on the expertise of scientific oceanographers in sound transmission (for submarine detection) and wave prediction (for amphibious assault). Furthermore, being a patron of fundamental research always seemed, thanks to the scientist's own perhaps unconscious image-building, to be cleaner and more worthy than sharpening torpedoes. The Navy says that, with the National Science Foundation, it assumed major federal responsibility for developing academic and institutional capability in ocean science research in the sixties. It claims to have been largely responsible for establishing ocean science programmes at Johns Hopkins University, Texas A. & M., Oregon State, and M.I.T. and to have expanded Rhode

Island and the University of Miami. Miami now spends a good deal of money saying that it is the oceanographic centre of the world (San Diego says the same and has been doing so for longer). Navy contracts have certainly done much to lift Scripps and Woods Hole out of the plus-fours and plankton net era. The Navy has its own in-house Laboratory programme and runs no fewer than 34 ships in its ocean science programme. In 1967, 18 academic and private institutions were engaged in Navy-sponsored work at levels of spending greater than 100,000 dollars a year. Six of them spend over a million Navy dollars a year.

Internal Navy politics have also reflected the rising tide of interest in oceanography. In 1961 the Hydrographer's title was changed to Oceanographer and the 1970 incumbent was of higher than usual rank, a rear-admiral. Paradoxically the Oceanographer of the Navy, Rear-Admiral O. D. Waters Jn. is a general service officer once specialised in gunnery. He has great charm and is presumably a good operator in corridors and committees. His function is management and you don't necessarily have to be au fait with copepods and acoustic transducers to pull together the fragmented unco-ordinated segments of U.S. Navy research. This, as a U.S. Navy report put it, 'is centered' on 13 separate activities.

The trouble is that oceanography is a sprawl anyway and even the perspective of defence is little help in making sense of it all. Oceanography's fundamental problem from the P.R. point of view is that it can never mount a project like the Apollo moon flights in which there is one identifiable goal that everyone can understand like walking about on the moon and coming back with bits of rock.

The U.S. Navy's view of the sea was sonorously stated in its film *Mission Oceanography*. 'The Oceans were once barriers. Now they are secret pathways to our shores.'

Anti-submarine warfare is thus the prime consideration when any particular project comes up. It so happens, that anti-submarine warfare is as good an excuse as a scientific oceanographer could wish for. If Admiral Waters is asked to define a national goal in oceanography he says, and it does not sound half so inspiring as sending a man to the Moon or to Mars, 'a complete understanding of the propagation of sound energy through the water'. If the U.S. Navy had such knowledge the oceans would no longer be a secret pathway to America's door. It would presuppose that America had a more or less complete knowledge of the topography and texture of the ocean bed, a complete catalogue of the sound characteristics of all animals including those that make up the elusive deep scattering layer, comprehensive predictive power in surface and sub-surface wave conditions, temperature distribution, salinity distribution, and atmospheric and tidal pressure distribution; America would also find useful a complete magnetic anomaly map of the world's oceans. With such knowledge, massive computers, continuous satellite, aerial and sonobuoy surveillance, and the co-operation of her allies it just might be possible to keep track of Russia's 400-odd submarines, except, perhaps, those lurking under the ice. The U.S. Navy would then know most of what there was to know about oceans except for heat flows, mineral distribution, and the structure of the bottom. Such knowledge would be a fair basis for a national ocean industry.

Therefore so far as ocean exploration is concerned the Navy has a carte-blanche. It can do what it likes, constrained only by cost, priorities, and the availability of suitable manpower.

The Navy has three Assistant Oceanographers. One of them deals with ocean science, another with ocean engineering and a third with oceanographic operations. Ocean engineering and development in a typical year like 1968 cost 100 million dollars. Oceanographic operations cost about the same and ocean science cost five million dollars.

As might be expected the ocean engineering programme envisages the most advanced technology of all. Oceanographic operations are comparatively conventional and would better be recognised by British readers under the heading hydrographic survey. The major output is charts and again, as might be expected of the Americans, many advances have been made in automating their preparation.

The ocean science programme however must be discussed here because it bears on the third great element in American oceanography, the universities and institutes. The ocean science programme may cost less than the other two but its prestige within the Navy is high. Of a group convened in 1956 to review undersea warfare over half came from the Navy's ocean science programme; the group's corporate thinking led to several of the current weapons systems. The brief is broad. OpNav Instruction No. 5450 165 says that ocean science is 'that effort research; development; and technical guidance in support of operations, to advance the knowledge of the physical/chemical/biological/geological nature of the world's oceans and their boundaries (surface and bottom).' It may not be English with its semicolons, slants and brackets, and paucity of verbs, but it certainly encourages the Navy's scientists to go where the spirit of scientific exploration takes them. Defence is however always looking over the shoulders of the men who are actually on the staff. There can be no other explanation for the Navy's failure to spot the significance of the striped reversals of the magnetic field patterns on the North Atlantic ridge. Such inhibitions are less strong when a Navy contract is given to a university or institute. Contract research has taken anything from 70 per cent upwards of total Navy ocean science funding.

The third and last major influence in United States policy will take time to show results but eventually it may prove more important than all the others, that is the sea grant programme.

The curious title is derived from the land grant colleges movement of the early years of the century. In this, individual states were encouraged by the federal authorities to set up educational establishments in the farming areas of the Mid-West. The technique was to provide grants of federal land and therefore income from rents to help out any local initiative in providing a usually vocationally orientated public education. As a result of the federal initiative land grant colleges grew up and provided an educated labour force which many economists see as being a primary influence on the vigorous development of the farming states.

The National Sea Grant College and Program Act became law on October 15th, 1966. The government hopes it can become a cornerstone of long-term investments to marine resource development. The programme places a special emphasis on developing regional capabilities to work on regionally orientated problems.

It has three objectives: it is supposed to provide more ocean engineers, it is supposed to push research into marine resources, and it is supposed to tell everybody about it and offer extension and advisory services.

It can work in two different ways. Firstly it can help existing academic institutions evolve into sea grant colleges which are then supposed to provide regional leadership in ocean research and development. Secondly it can support projects which involve specialised scientific or engineering research, or even education advisory or training activities. The National Science Foundation is supposed to do the initiating and supporting of sea grant programmes.

The emphasis is on strictly short-term prospects capable of reasonable financial return: food from the sea, environmental forecasting, continental shelf exploitation and multiple use of the sea coast. The last refers to the need to reconcile competing uses of the coast, amenity, fishing, dredging, polluting, and recreational sailing and boating.

The fundamental notion is pump priming. 'To mobilise collective resources, foster local channels of communication and establish new patterns of collaboration beyond the limits of any single sea grant institution ... industrial requirements can help to shape the direction of the program and indeed the NSF looks to industry for suggestions and ideas ...' Available qualified manpower may be expected to double within a decade, 300 or more a year of graduate ocean engineers and perhaps 1,000 a year or more of ocean technicians.

Lest anyone lacking in imagination should fail to see quite what marine resources have got to do with such routine educational chores as teaching people domestic science, law, or journalism the document 'Marine Science Affairs' January 1967 (Report of the President to Congress on marine resources) spells it out at length. Federal money up to a limit of two-thirds of the total will help a state institution's school of law if the students buckle down to studying the legal aspects of marine resource development. Even a school of domestic science or home economics can get money from federal coffers if it deals with the 'home utilisation of marine foods', that is presumably American for cooking fish and cleaning mussels. If anyone wants the money they have to take it seriously, a host institution must adopt the sea grant mission as a major goal and a full-time programme director or co-ordinator will be regarded as earnest of such intentions.

Sea grant project support is intended to advance know-how most of all at the stage where scientific discovery first blends into a possible social application, but where practical economic benefits are still uncertain.

In its first full year, 1968, total funding for the sea grant programme came to five million dollars for 27 grants at Oregon State, Rhode Island, Washington, Hawaii, Texas and Wisconsin Universities. In 1970 the money available doubled that available in 1968.

As an example of the thoroughness with which the Americans go about this sort of vocational education one may take the curriculum for 'ocean technicians'. This is for use in junior colleges giving two-year courses and is below first degree standard. There is no doubt that any man emerging with success from this sort of course would be a most excellent hand in any sort of marine activity. An ocean technician will have survived three courses of technical English in which he will have even conducted meetings and conferences. His technical mathematics will embrace differential and integral calculus and his technical physics will include electron theory and properties of matter. Navigation and seamanship sounds as if it is up to the sort of standard required for junior officers in the Royal Navy and includes celestial navigation. Making charts also features. That is only half of the ambitious curriculum, however; oceanography, fishing and marine biology are also introduced with emphasis on the practical. On top of that, the budding technician is to be capable of stripping a diesel, petrol engine or pump and will be trained to be handy with rigging and anchor laying and keeping a marine refrigeration system going. Lastly this salty paragon will be able to type and use a desk calculator! The only thing he cannot do, dare one mention it, is dive and swim but that is probably regarded like car driving as something every boy knows about anyway.

If the sea grant programme goes on to fulfil its objectives America will soon have the most awesome capability in marine activities. The momentum provided by such a solidly based labour force will be irresistible and if there is wealth in the oceans then the United States will get it.

Those three factors then have played the most important parts in generating the new big science of oceanography. To recapitulate, they are first, the U.S. Navy's need to maintain mastery of the sea; second, the advanced technology companies' wish to deploy their skills in new fields related to aerospace; and third, the skilful deployment of federal money in the education system.

Together they have raised the recurrent spectre for the rest of the world, a runaway American lead in yet another area of advanced technology. The rest of the world has responded after its fashion. The British, who were, indeed possibly still are in 1970, slightly ahead of the United States in the pure science side of oceanography, reacted with the slow, mumbling suspiciousness of a punchy old boxer. The French, led by the glamour boy of oceanography, Jacques Cousteau, established a beatuiful formal structure and a National Centre. The Germans and the Japanese in their turn are watching the Americans closely and in 1970 there was a feeling of relief that at least things were not happening as fast as had once seemed likely. The Russians who have militarily as much at stake as the Americans have said as little as possible.

10

Where is Everybody?

THE AMERICANS ARE A COMPETITIVE NATION AT LEAST IN TECHNOLOGY AND sport, if not in social welfare. One of the reasons that the American supersonic transport was persevered with so long is that the Anglo-French Concorde, which is scarcely less controversial, is still roaring on through its flight tests. If the French and the British gave it up, it is quite probable that America's S.S.T. would be buried for good. Similarly in oceanography if all the European nations together had reacted to what might be called the American challenge, the United States would be swimming along much more strongly in the race to keep ahead.

In the mid-sixties, when the British were asked by their own internal critics what was their oceanographic policy, it would be fair to say that they did not know whether they had one at all. They had not regarded ocean activities like fishing as having anything to do with continental shelf exploration or research voyages to distant parts of the world. Furthermore they were innately suspicious of the American style, the nomination of the ocean as man's last frontier, a great scientific and technological arena in which great deeds would be done by the joint will of science, government, and industry. The French, however, seem to be more susceptible to this sort of challenge. The primary reason is French patriotism, a profound belief that France occupies a place in the world pecking order which is less exalted than, intrinsically, it ought to be. One may point to the French efforts to get a satellite into space and to the splendorous glittering style of her national exhibits at aerospace shows. At first sight it often seems a bit pathetic but other nations tend to forget that down where the work is done there are shrewd and thrifty businessmen who see money just as clearly as the government sees *gloire*. M. Marcel Dassault is a good example from the aerospace sector. The French government's backing of oceanography can provide many entrepreneurs with opportunity. In oceanography the French also had the asset of Cmdr. Jacques Yves Cousteau who has been an indefatigable propagandist for all sorts of marine activity under the sea; he made oceanography glamorous and allowed the government to spend taxpayers' money more easily. France was able therefore to respond to the American challenge more consciously than most other European nations and she did so in a way which is comparatively easy to describe, because it is so logical.

The central notion was to invent a body called C.N.E.X.O., the *Centre National pour L'Exploitation des Océans*. C.N.E.X.O. was from the beginning a centre for exploitation not for research. It has been going since 1968 and has about 60 people working for it; what it mainly does is to plan. It has its own funds to deploy in its chosen areas. The list of areas follows a well-worn trail; first comes fishing or living resources, then minerals and oil, pollution and the effect of the ocean on climatic conditions.

Contracts have been placed with various universities for marine biology studies and for mapping the continental shelf. In the field of instrumentation and navigation it has given contracts to the private firm COMEX (q.v.). Like most bodies whose future depends on the vagaries of politics it has taken the precaution of plunging into bricks and mortar by building a centre of oceanography at Brest which should be ready in 1975. By then it should house 400 people. The level of funding for outside projects is still low but the planning effort is considerable. It is easy to scoff at planning bureaucrats but at the present stage in oceanography intensive paper studies are probably well justified. C.N.E.X.O. disposes of one third (58 million francs) of total public expenditure on oceanography.

Plainly the offshore oil business has the biggest flow of money and French national oceanographic organisations are well placed to tap it for the benefit of France. For one thing French oil companies have always been patriotic, they have always co-operated with the government's French Oil Institute (I.F.P.) founded in 1944. This large organisation with a budget of

15 million pounds has a staff of 1,500 people and a definite programme of marine research. The objective is to get oil from some place other than the Middle East and this means in large part the continental shelf in the Bay of Biscay. The marine programme is helped by a slice of the government's petrol tax. The I.F.P. has thus given certain sectors of oceanography a big stimulus and has ensured that the oil industry takes an interest.

This is in plain contrast to the British experience where large national oil companies have made no effort to support specifically national oceanographic enterprises. Indeed British Petroleum, in which the British Government itself has a huge share-holding, recently chose to link up with the French company COMEX on deciding to participate in an oceanographic development company.

The I.F.P. and C.N.E.X.O. between them embarked on the building of a high performance submarine, the *Argyronete*. This craft was designed from the first to work on oil well heads and to lock divers in and out under the sea. Its maximum diving depth is 600 metres. France also has access to the Navy's bathyscaphe *Archimède* and Cousteau's Diving Saucer *SP 3,000*. The *Archimède* will dive to 11,000 metres and the *SP 3,000*, as the name implies, to 3,000 metres.

There are two other lively and energetic organisations which help French oceanographic capability. One is commercial, the COMEX company already mentioned, the other is so peculiar as to be just French. It is the Centre of Advanced Marine Studies (CEMA) which is built around the person of its Director, Commander Cousteau.

CEMA is supported by whatever funds it can get but Cousteau's T.V. films now contribute a good deal to its running. Cousteau himself is a sort of world citizen of diving and enjoys the confidence of everybody from the U.S. Navy and Westinghouse to his own government. In 1969 when U.S. companies had regretfully put their myriad research submarines on the stocks and hard-faced finance directors were cutting back, Cousteau somehow talked a dozen of them back into the water for some aqueous epic in which all the submarines cruised about together. Most of them were in deadly rivalry for what few contracts there were around and the fact they all docilely submerged together at the Frenchman's command was as fine a tribute to his standing as anyone could wish for.

Cousteau helped Westinghouse into oceanography and built one of their submarines – *Deep Star*. He was the leader in the underwater house business and dramatised the idea of man living under the sea on a long-term basis. It is easy to call many of the things Cousteau does publicity stunts but in the context of the rise of oceanography many would prefer to see his work as sophisticated and highly effective propaganda as well as solidly creative engineering.

In every country there are one or two companies in oceanography that actually make a profit even at the early stage; one of the most notable in France is COMEX, *Compagnie Maritime d'Experitises* based in Marseilles. COMEX has grown rich and big by making sure it stayed near the big money in the oil industry. It has an impressive diving capability and is pressing at the 1,000 foot barrier in experimental situations. Like Lockheed and Ocean Science, it is building a dry system for oil well head maintenance at atmospheric pressure. Most people think it is an impressive operation, even the Americans. It has underwater welding technology and is building instrumented buoys for the oil industry.

As a nation the French have gone about oceanography in a measured, determined and economical way. The virtue of the French effort is that one can see what they are doing, and gain some idea of what they are likely to achieve. Furthermore, French participation in the North Sea Oil search has given many of the big American contractors something to think about. The trouble is that the absolute magnitude of the French effort is puny. In 1969 excluding Navy financing, it was of the order of no more than 20 million pounds.

The United States budget in the same year for broadly comparable activities was 300 million pounds or more. Even if all the budgets of the European nations were lumped together they would not amount to a quarter of that sum, and when duplication is allowed for, the effective expenditure might not amount to one sixth of it. Germany's public expenditure was three million pounds, one hundredth of the U.S. expenditure. British expenditure was 14 million pounds.

Even allowing for widespread duplication within America and considerable inefficiencies in the disposal of funds, the number of effective man hours expended by the United States must be at least five times that of the whole of Europe. In the absence of serious competition, therefore, the American government has increasingly had to look at civil oceanography as a straightforward business proposition. Unlike the space competition with Russia, America has not been able to indulge in state sponsored technological P.T. The Americans have used space as a vast gymnasium to flex and improve their technological muscles but except in one particular they have had no motive to do the same in the deep ocean. The one exception may possibly be a vital one in the long term. It is the behaviour of sound in the sea; sound transmission takes the place of light, radar and radio on land. The Russian military oceanographic effort is believed to be slightly larger than that of the United States and seems to be almost entirely devoted to this subject. It probably ranks as a 'national goal' in Russia as it does in the United States.

It is very difficult to give any realistic assessment of Russian oceanographic policy because of that nation's mania for secrecy. It is known that Russia has a big interest in the magnetic contours of the ocean bed and has a specially built ship the *Zarya* which is non-magnetic, it contains no steel at all. Indeed it is thought to be the only non-magnetic ship in existence. On the instrumentation side Russia is hampered by the poor quality of her electronics components industry. The Russians spend a lot of effort at 'buoy bagging' from the Americans. It can't be called stealing since buoys in the ocean have no legal rights and can be salvaged by anyone with a mind to do so. The Americans for their part bag Russian buoys, no doubt to find out how much copying goes on. Another feature of Russian oceanography is a touch of Lysenkoism. Just as Lysenko denied modern genetics so did one element in Russian oceanography deny the significance of the mid-ocean ridge. From the point of view of civil oceanography it is still too difficult to predict how much spin-off will come from defence efforts. In any case Russia's unexploited natural wealth means that she has far less reason to attempt ocean mining than Japan or Britain or America. Furthermore her vast fishing fleet staffed by a badly paid working class means that fish protein does not yet have to come from fish farming, and Russia's coastline is an unhelpful one fronting onto icebound Arctic waters or the confined Baltic and Black Seas. Nevertheless, as Russian naval power expands, so will her oceanographic activities. Encouragingly she has shown a willingness to join in international ventures in pure science and, with the French, in such applied science ventures as designing deep ocean corers.

Of the other advanced countries Britain, Germany, Canada and Japan have significant programmes. Canadian figures for 1967 show an annual expenditure in basic and applied research amounting to over 15 million pounds. The effort is spread over most fields though dredging seems to have little importance. Canada had not by 1969 designed any programme on a national basis but it can only be a matter of time before she does so.

Pollution in the Great Lakes is one good reason, likely pollution along the Arctic coast if oil tankers take the North-West passage is another. Furthermore the Canadian continental shelf is 40 per cent as large as the total area of Canada and so represents a huge resource. Oil on the north slope of Alaska has already pointed the way to the development of the adjacent Canadian areas.

Japan is starting slowly but there are good reasons why Japan can be expected to become a notable power in oceanography. She of all nations must look to ocean mining as a source of most of the minerals which a modern nation requires. Her national product will soon rank second in the world after America's but, unlike the rich continent of North America, Japan has but a small land area poorly supplied with minerals. Nor has Japan any oil, indeed her economy could barely function without large raw material imports. In search of offshore oil she has already begun seismic exploration to the east of Hokkaido, the northernmost island, and to the west of Sendai in Honshu. At the other end of Honshu, the Japanese main island, Shell Mitsubishi was due to drill near Yamaguchi in 1970. Japan has already begun work on the problem of extracting metals from manganese nodules and has processed two tons of nodules brought back from offshore California. Deep drilling techniques are under development and 11 million pounds has been set aside for the construction of the submerged drilling rig (see ch. 11). Private industrial companies have started on their own, Komatsu's underwater bulldozer is a good example.

Japan sees more clearly than most that in modern industrial societies economic advance can proceed by the creation of big science. Such ventures require close co-operation between government, industry and academic scientists and Japan is better placed to put what they call 'systems industries' into action than other nations where traditions of independence are more jealously preserved. Systems industries the Japanese believe are the key to the future in advanced technology. Many specialist companies each with their own motivation and traditions must work together to realise joint projects. They must not be ad hoc consortia knocked together for the occasion but really co-operative friendly companies pulling together to achieve some new enterprise of economic significance. Japanese companies have a long tradition of co-operating with the government, and scientists in Japanese universities have a long tradition of co-operating with both government and industry, all in the cause of advancing the prosperity and standing of the Japanese nation. Clearly a project like a submerged drilling rig will require all sorts of concerted action and a late delivery or a failure on the critical path will prove cripplingly expensive.

Britain is curiously like Japan in many geographical features. She has now no mineral resources worth mentioning (for example Cornish tin will probably never amount to more than 10 per cent of Britain's needs) and like Japan she has to import all her oil. If ever a nation needed to invest in oceanographic projects, it is Britain. Her fishing industry has a private balance of payments deficit of nearly 100 million pounds, she imports millions of dollars worth of petroleum products and essential metals. She has a shipbuilding industry which the politicians at least regard as important, she has a great naval tradition and a huge shipping industry. Her standing in the science side of the great ocean business is still of equal ranking to that of the United States. Of the smaller nations in the world she has most to gain from the oceans, and the best qualified manpower for doing the job.

What does Britain do? A policy white paper was published in 1969 and reported on marine science and technology. The Ministry of Technology had a government aviation man in charge of it all, a world pioneer in the techniques of blind landing aircraft in fog. The white paper was fairly smug and confirmed that (with the implication that it was already widely believed) 'the United Kingdom's work in marine science and technology is in general well balanced, and that there are no major fields in which our efforts are significantly inadequate.' The general attitude in the British government is that Britain's annual expenditure of 13 or 14 million pounds in a year is in the same ratio to her gross national product as is America's 300 or 400 million pounds invested in the same field. More sophisticated arguments are not admitted, though to the uninitiated they may look more relevant. For example the total U.S. product from the sea (1,100 million pounds in 1967) is only twice that of Britain but it commands an investment over 20 times greater. Such arguments are regarded as special pleading in spite of the fact that the rate of investment in general in Britain is half what it is in Japan and consistently lower than in any European country.

If one assumes that investment creates wealth and therefore income, one must conclude that Britain's days as a major maritime influence are numbered. It is just possible that British investment is wiser or less foolish than that of other countries. In particular American expenditure may be demonstrably uncommercial in the short term or even in the medium term. There are many 'told you so' sentiments expressed at the spectacle of America's many research submarines on the beach. The British like to think they are as rapiers to the American bludgeons. This is particularly true in pure science where it is said a little high-class thinking on the part of one man can achieve results the Americans get with 20 Ph.D.s.

Unfortunately even in science the practice of doing things on the cheap may cease to pay dividends. In Britain for example no-one can get enough ship-time because there aren't enough ships. In America single institutions like Wood's Hole have more research ships of their own than the whole of Britain. The conditions for yet another brain drain have matured. However, marine science did not suffer in the general science cutbacks of recent years when most expenditure was held to eight per cent, in oceanography, expenditure is officially allowed to grow at a higher than usual rate at 12 per cent.

Morale at Britain's National Institute of Oceanography (N.I.O.) which is a good way from the sea at Godalming, Surrey, is therefore still high. It basks in the reputations of men like the Director, Sir George Deacon, C.B.E., F.R.S. and researchers like John Swallow, M. S. Longuet-Higgins and Anthony Laughton. Similarly in the universities great names are still active like Sir Alistair Hardy who invented the plankton sampler towed behind merchant ships, and Sir Edward Bullard a pupil of Rutherford who specialised in geophysics. In the field of science controlled expansion goes steadily on thanks to expenditure which still rises faster than inflation diminishes it. The acceptance of the need for higher expenditure clearly implies a shift in a nation's attitudes towards a particular field of activity, in this case oceanography. But attitudes in the minds of individuals cannot be switched so easily and in Britain the division between pure science on the one hand and applied science and industry on the other is very profound. Recently a merchant banker's survey showed that even in an institution like the Imperial College of Science and Technology in London very few scientists had any desire to make money out of their own innovations and discoveries. Industrialists and entrepreneurs for their part find they have little in common with scientists, and communication across the divide is poor. These attitudes are slowly changing; it has been made clear by the various research councils which dispense government money to university scientists that in future projects with some industrial application will be more favourably looked upon than those which have none.

In the world of pure science there is a feeling that support is, if not marvellous, as good as can be expected. The situation in industry is far less comfortable. It is all very well making scientists think of industrial applications but if industrialists couldn't care less the crossover from scientific discoveries into commercial hardware and profitable activity will never happen. However, several adroit propagandists have been at work and their activities have been as interesting to watch as that of some of the American operators. One of the leading lights is Rear-Admiral Sir Edmund Irving recently retired Hydrographer of the Navy, an old sailor who radiates a great and fatherly charm which conceals considerable toughness. Getting to the top of the Hydrographic Survey branch of the Royal Navy demands great strength of character. The Royal Navy works its survey ships incredibly hard and they get a great deal of sea-time in exhausting places like the Persian Gulf. Officers have been known to go definitely peculiar under the strain. The Hydrographic Survey Service consumes six million pounds a year out of Britain's total so its boss is the largest single consumer of oceanographic money in Britain. When he left the Navy, Sir Edmund Irving became the chairman of the Society for Underwater Technology (S.U.T.) which is the principal source of political pressure for oceanography in Britain.

It was the S.U.T. which sponsored the Brighton conference 'Oceanology International '69'. It took place alongside the International Oceanological Equipment and Services Exhibition arranged by B.P.S. Exhibitions Ltd., and it drew lavish exhibits from companies all over the world. The week-long event in February 1969 had the same effect as the American monthly magazines, it substantiated the ocean industry, it made it seem more real. Journalists, Members of Parliament, businessmen and civil servants all took the chance of a day away from the office by taking the train down from London to the well-known seaside resort. Oceanography got more column inches that week in the British Press than it would normally get in two or even three years. The Minister of Technology had to say something. It was all about partnership between industry, government and universities. He also went round the exhibition. Sir Edmund Irving got in his piece about the neglect of ocean technology. The journalists skimmed John Mero's book about deep sea minerals and gazed at maps of the Red Sea brine sinks, they clustered round bits of surprising hardware like the diver's 'motorbike' and duly reported that sea research costs might soon rival space projects, at least they might for Russia and

America. The Sealab III experiment began in California: men would live for a month at 600 feet. There was a model of it in Brighton and a state board giving details of progress. The almost immediate fatal accident gave an added though very unwelcome frisson to the proceedings. The pile of published conference papers stood over one foot high and formed an invaluable source of material for the many British companies who began to ask their planners to take a look at oceanography. In terms of pressure politics it was a triumph.

Another of the men behind it, N. C. Fleming of the National Institute of Oceanography was already experienced in the subtle arts. He was involved in the preparation of a secret report for six companies, Costain, Unilever, B.P., Rio Tinto Zinc, I.C.I. and Hawker Siddeley. This report was the source of all sorts of dark allusions and rumours to such an extent that competitive companies began to feel uncomfortable at the prospect of a rival stealing a march. Companies like the British Aircraft Corporation thereupon began their own investigations. The result was that the Commercial Oceanology Study Group's report had a far wider impact than its six sponsors ever intended.

Another element played a notable part in selling oceanography in Britain, though it was and is the subject of much cynical comment – the Atomic Energy Authority (A.E.A.). Its Harwell laboratories had run out of work and since Britain's civil atom power-station policy was in disarray and no-one knew what to do with the research teams the A.E.A. started looking for work itself. In some ways their situation was like that of the U.S. aerospace companies, they had too much talent and not enough to do. Two talents were especially appropriate to oceanography, one was the remote control skills developed for handling radioactive materials, the other was know-how in the field of corrosion-resistant materials. In 1967 Harwell decided to educate itself and in its background brief announced that 'The Ministry of Technology has therefore asked the Authority to provide an occasion on which scientists and technologists from industry and Government can exchange views.' The programme of the conference was arranged by a committee on which the Society for Underwater Technology and the Commercial Oceanology Study Group were represented. Those attending the conference included scientists from many companies and government laboratories, more than 40 papers were given and the representatives of the money bags, The Ministry of Technology, listened carefully but stood well back.

In spite of these multilateral blandishments policy makers have not responded. All that has happened has been the creation of another committee, the C.M.T. or Committee on Marine Technology and a lot more emphasis on co-ordination. The C.M.T. is not C.N.E.X.O., it has a staff of only four and is far too small. Government policy in a country where nearly half the domestic product is handled by state-owned industries is unavoidably a vital factor in the development of new enterprises. This is why so many efforts have been made to influence the government's attitudes.

Consider the British dredging industry, this was once one of the world's best but it is now completely overshadowed by the Dutch. The reasons are partly the decline of the Clyde shipbuilding industry in spite of injections of the sort of taxpayers' money that would set the oceanographers rolling in their boats. But a more important reason is that public policy has never given the dredging industry anything to do. For example, to get gravel for concrete near London it is customary to bulldoze through acres of farmland and forest to get at the Thames gravel terraces. One company recently bought up a stately home's associated estate with a mature ornamental landscape of high recreational and historical value inside the London Green Belt just to get at the sand and gravel underneath. Over the next 10 years upwards of 50,000 acres of agricultural or recreational land are due for this sort of treatment. Public policy bears on this situation in two ways. Firstly, it can be much tougher in its opposition to the vandalisation of the countryside by the extractive industries so that there is more pressure to dredge aggregate offshore. Secondly, it can ease the opposition to sea dredging which comes from coastal local authorities and ports. It can do this by sponsoring programmes of research to find out the likely effects of dredging aggregate from what are, in the North Sea, the submerged river channels of the Rhine Delta. So far only 10 per cent of total British requirements come from the sea.

To be fair the government is moving in this direction but very late in the day. Indeed for years it has solemnly collected one shilling a ton from Dutch dredgers taking gravel from British waters for use in the building of Europoort, and Rotterdam, a vast modern port complex which has now dwarfed the Port of London. The further consequences of this bumbling policy are that the possible development of Foulness, a Thames estuary site, for London's third airport, is likely to be entrusted to foreign contract dredger companies. Meanwhile no doubt, there will be further subsidies to prop up declining British dredger builders who complain of the lack of a large home market.

British oceanographic activities are fragmented under various ministries and bodies, a short list includes the Ministry of Agriculture, Fisheries and Food, the National Ports Council, the Ministry of Technology, the Ministry of Defence, the Natural Environment Research Council, the Ministry of Agriculture and Fisheries for Scotland and the Ministry of Transport. Naturally this gives scope for many interdepartmental co-ordinating committees and multiplies the number of toes to be trodden on when anything is to be done. There are many civil servants who would object to the use of the word 'fragmented'. What has sea bed geology got to do with pollution control or defence? It depends how you look at it, of course, they do have something to do with each other or there would not be so much effort at co-ordinating. Lumping them together and letting a British variety of the 'wet NASA' now established in America grow up within the government machine would obviously sanction an oceanographic bandwagon. One cannot help feeling that this is what Whitehall wishes to prevent.

The British government is not the only one to feel that way; plainly the American government also feels that a National Oceanic and Atmospheric Agency (N.O.A.A.) along the lines suggested by the Statton Commission would be too strong a source of pressure. It has finally sanctioned its creation but is keeping it short of money. The proposed International Decade of Ocean Exploration has become moribund for similar reasons. In both countries the fundamental problems are economic. In Britain the economy is basically weak and growing only very slowly. An adventure is the last thing the British Treasury wants when large sums are continually being demanded for wage rises in the public sector along with government promises of tax cuts. Many would see Concorde as an existing example of an adventure consuming colossal sums of money, appellants for government support frequently ask plaintively for a landing wheel or the tail fin. But as Concorde began as an adventure before oceanography built its bandwagon, it seems likely that no other bandwagon will be allowed to start until Concorde is over. Ironically, however, the likelihood of a prototype Concorde falling into the sea is one reason that the government is likely to keep at least one research submarine operational. It would be needed to pick up the bits.

The United States too has its money-burning adventures, the prime one being the Vietnam War. In addition, the economists of the Chicago School have persuaded the government that the printing of banknotes should be restricted in order to combat inflation. Therefore most innovation and high risk enterprise is becoming progressively more difficult and major companies are more reluctant than ever to risk their own money. However the lack of a 'wet NASA', in the United States has been of far less importance than the lack of a similar centre of pressure in Britain. The reason is the sheer scale of American budgets in general and the cash flow from offshore oil.

The private sector in Britain, now dubbed 'Hawks' by the Press presumably to distinguish them from the Treasury 'Doves', is extremely weary of the flood of words and the lack of action and leadership from the government. But apparently the French feel the same way and the prime critic is, as might be expected, Cmdr. Cousteau. Therefore a European enterprise has been launched and called, inevitably and sonorously, 'Eurocean'. It will consult, fix contracts and continue to berate·the public sectors. But nations and politicians being the way they are, the great ocean business is unlikely to leap forward as fast as its supporters would wish. It's rather like those bicycle races where they all pedal along slowly for ages until someone suddenly makes a break. A reasonable bet in the circumstances would be that Japan will make the break and America will win.

11

The Drilling Must Go On

WHATEVER THE POLICY MAKERS DECIDE, OCEAN EXPLORATION GETS STEADY encouragement from the oil industry's need to drill in ever more difficult environments. The thing about drilling that always causes the greatest surprise is the scale. It must be, apart from cable and pipe laying, the world's most elongated technology. When the Mohole project was first mooted one of the authors was working on a television science programme, and decided to do a studio story about it. In the fifties, British television took itself very seriously and it was decided to match the scale of the Mohole exactly with a model and then interview the British Petroleum (B.P.) geologist, Dr. Tom Gaskell, who was one of the instigators of the project.

The Moho is the boundary between the earth's crust and the mantle. It is at its shallowest depth in the ocean where it may only be 10 kilometres from the surface of the sea, under only five kilometres of solid rock and sediment. The width of the top of the world's deepest producing oil well at that time was only 20 inches. A 30,000 foot hole, scaled down to the height of the studio might be represented by a 30 foot long piece of wire. To preserve the scale its width would have had to be a fiftieth of an inch or the thickness of a fine pen stroke. It was decided that such fine wire would be both invisible and certain to break in the studio. Something much thicker was needed. The item therefore began with an apology that the bit of thick piano wire which disappeared up into the lighting gantries in the studio roof was far too thick to represent the Mohole. One of the participants ran it between his fingers and said, 'Well, if that's to scale it's not an oil well, it's a mine shaft!' In the event it was still too thin to be seen by the T.V. cameras and the item proceeded with ghostly references to a totally invisible model.

Conventional drilling depends on rotating that extraordinarily long and thin piece of 'drill string' from the top in order to transmit twist to a drilling bit at the bottom. Heavy collars sit on the drill bit so that it can bear down on the rock with considerable weight if the driller cares to let it. If a 50,000 foot hole were to be attempted, however, the weight of the conventional drill string would be greater than its own tensile strength and even if made of high tensile steel it would have to be handled gingerly when being run in and out. Clearly boring deep holes is expensive and risky.

Most holes drilled on land rarely penetrate beyond 20,000 feet but even then they cost a lot of money. Ten years ago when the Mohole was a serious possibility offshore drilling costs were 50 dollars a foot. Most of the wells drilled into the North Sea, for example, cost at least 1·5 million dollars each, and in general nine dry ones are drilled for every productive one. The flow of money needed to keep going in the offshore gas and oil business is therefore colossal. The total capital cost for exploring and developing a 500 million cubic feet per day field on a 15-year basis may be 100 million dollars. A single drilling rig like B.P.'s Sea Quest costs nearly nine million dollars and, displacing 15,000 tons, is equivalent to a medium-sized cargo ship. There may be up to 10 such rigs operating in an area like the North Sea at any one time and in 1969 the world total was around 200 rigs.

The American domestic offshore oil industry invests a billion dollars every year excluding lease payments which cost as much again. These expenditures are expected to grow at the rate of 18 per cent annually according to the Stratton Report. By the end of the seventies, one third of the world's oil supply will come from undersea production. Oil and gas have been discovered way beyond the 500-feet water depth which is broadly speaking the present limit of commercial activity. The deepest recorded indications came from a scientific drilling at 12,000 feet on the Sigsbee Knolls in the Gulf of Mexico.

For oceanography, offshore oil and gas therefore provides a giant swirl of money, far far more than from any other source except shipping. Anything which helps the drilling men to drill even one per cent more efficiently can look forward to a fair amount of support from the industry and, for more long-term projects, from governments.

In Britain, government money has already been provided for an underwater bottom crawling vehicle and various other bits of advanced technology entirely on the basis of market research into the demand for diving services on drilling rigs. Devices costing large sums of money can be justified on the basis of giving divers a bit of extra help or time on the bottom when clearing trouble on a rig. It is the same sort of economics that rule the film world. When a scene is being shot and the director wants a picturesque old dinghy to be rowed in the background behind Sophia Loren, it is cheaper to buy the boat on the spot for 1,000 dollars than to haggle over whether it is available for hire that afternoon or not. If the whole production is held up only one hour it will cost 1,000 dollars anyway. It is the same with drilling rigs; if they are not drilling they are eating money at a rate which is of the same order of magnitude as that for film stars and film crews, up to 10,000 dollars per day. Deep heroic wells cost almost as much as Hollywood's big heroic films.

The techniques of drilling have changed little in principle since the earliest days of oil prospecting. The Russians have shown enthusiasm for breaking away from the conventional, but nothing they have produced has yet achieved commercial success. One post-war effort has been the exploitation of the turbo drill. The idea of this is that since some fluid or other has to be pumped down the hole to wash the chippings back up to surface, it is sensible to use the fluid to drive a turbine at the bottom of the hole. The rotation thus arranged drives the drill bit. One problem, however, is that fluid for lifting chippings is not necessarily good for driving turbines. Another technique is to drive the drill with a slim electric motor at the bottom of the hole, but some way has to be found to stop the motor rotating instead of the drill. The latest example of Russian determination to drill by some method different from the conventional American style is the use of rocket blast to fragment the rocks.

The only method likely to have a future seems to be the turbo drill which has received the attention of jet pioneer Sir Frank Whittle and Bristol Siddeley Engines in Britain. Making improvements in drilling runs up against the same problems that novel motor car engines always encounter in trying to displace the conventional piston engine. The conventional car engine has been under steady development for years, everybody knows how to work it and repair it. No doubt if someone had lavished the same development money on turbines or steam cars, modern cars would be far better than the petrol-engined ones, but no one did, and it seems too late to start now unless some really capital advantage can be gained. In drilling however, there is one capital defect in the conventional techniques, that is the enormous time wasted on 'round trips'. A round trip is the business of pulling up the worn-out drill head, changing it, and then lowering it to the bottom of the well again. A well in Texas that went to 23,880 feet took 228 days to drill, but one fifth of those were consumed in round trips. If turbo drills were successful it might be possible simply to wind them in and out of the hole on a cable. In that case round trip time could be reduced by an enormous factor and even if a turbo drill were not competitive otherwise that one advantage might make all the difference.

How does conventional drilling work? It is rather important to understand the technology because attending to its whims is, as we have said, one of the surest sources of finance for a wide range of oceanographic hardware. It is not difficult to understand since it is exactly what any ordinary handyman would suggest if the problem was put to him to solve today. In fact except for the details of the drill head, the technology is almost as old and in some ways almost as charming as steam railways.

The point of the whole business is to rotate a drilling bit and push it with a controlled force into the rock. The push is easy, it comes from the weight of the pipe joined to the drill head. The pipe transmits the rotary motion needed to twist the drill bit, and it has to be very tough not only because it needs to transmit twist (torque) but because it has to hold up its own weight. On top of all this it has to withstand high pressures of mud being pumped down inside it and its own abrasion on the inside walls of the hole. Drill pipe comes in 30 foot lengths and a drill string is composed of many such lengths joined together.

Putting twist onto the drill string could be easily achieved if the pipes were, say, square in section. However this would not only be expensive, because square pipes are harder to make than cylindrical ones, but it would make for much more friction down in the well itself. The way round this is to join a square section of pipe into the string at the surface and rotate it, then just before it disappears down the hole, hoist it up and replace it with

THE ANATOMY OF A DRILLING RIG

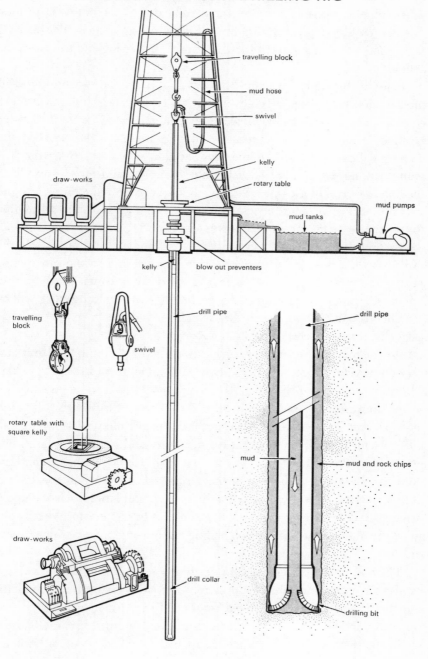

a cylindrical section of pipe. The top part of the drill string is then again made from square pipe until once again it is about to disappear. The square pipe is called a 'kelly' and it is rotated by being allowed to slide through a square hole of similar size cut in the centre of a circular 'drilling table'. This rotary table turns the kelly in the drill string and this turns the drill bit.

Hoisting the kelly up again before it disappears and substituting drill pipe naturally requires a crane or derrick to lift and hold the drilling column. The name derrick incidentally is supposed to be derived from Derrick, a dexterous seventeenth-century hangman who worked the Tyburn gallows in London in the days of public executions. Since the drill string hanging from the derrick has both to rotate and move down into the hole, a swivel block and 'travelling' block are required to hook it to the steel frame. The average derrick is up to 150 feet high and allows drill pipe to be joined on three or four lengths at a time during round trips.

By measuring the tension in the wires of the travelling block the driller can tell how much weight is actually bearing on the bit, by adjusting the weight the bit can rotate at the optimum rate. Furthermore, when the string is being withdrawn the first sign of its getting stuck will be a sudden rise in tension. The driller spends most of his time watching the large dial which registers tension or weight.

Power for these operations of hoisting the drill pipe in and out and rotating it via the table and the square kelly comes from the draw-works. This is basically a winch which, like the railways, used to be powered by steam. Indeed the parallel with the railways goes further, for although diesel electric engines are usual on modern rigs the old drillers would still rather have steam. They say that for flexibility and important things like producing maximum torque on starting steam cannot be beaten and indeed there are said to be many steam rigs still operating in the United States. So far there is no record of a steam rig preservation society and old drillers are denied the pleasure which came so frequently to old railwaymen, the sight of a steam engine chuffing up to rescue a broken-down new diesel.

There is, however, one vital bit of drilling technology that the average handyman, inventing from first principles, would probably never think of, and that is the crucial role played by mud. Early drillers discovered that water in the hole made the drill turn more easily and that the muddier the water the easier everything became. Such is the progress of technology that mud engineers now have degrees in chemistry and write learned papers on the behaviour of what may be as many as 20 different additives. Several men have devoted a profitable lifetime to the study and development of drilling mud.

There are several reasons for circulating mud during drilling. It is pumped down inside the drill pipe and escapes at the drill bit returning up the annular space between the outside of the drill pipe and the walls of the hole. One function is to cool the drill bit, another is to help break up the fragments and lubricate the cutting, yet another is to transport the chippings to the surface. A very important function is to counterbalance the pressure set up by water, oil or gas liberated from the rocks penetrated by the drill. To do this more efficiently, heavy minerals like barytes are added to the mud. Additives, of course, may affect the viscosity of the mud, this in turn affects the drilling rate and thus the rate of flow of money out of the oil company's pockets.

The only other item of fundamental importance is casing. Casing is used to prevent the cave-in of the walls not only in permanent wells which actually produce oil or gas but even in holes still being drilled. The drill string is withdrawn and steel pipe of slightly smaller diameter than the hole is run out. Thirty foot lengths of it screw together as with drill pipe. Cement is then pumped in instead of the drilling mud. The cement squeezes up between the casing and the walls of the hole and sets. It is driven down the inside of the casing by a cement plug, itself pushed by drilling mud. When the plug reaches the bottom and everything has set, the drill string goes in again, drills through the cement plug and carries on down into new rock in the same manner as before.

The top section of casing allows the blow-out preventers (B.O.P.'s) to be fitted to a solid anchorage. These are simply extremely robust steel valves capable of withstanding pressures of say 5,000 pounds per square inch. If oil or gas under pressure is struck they stop it blowing out and wrecking the rig.

One last trick of the land driller may be mentioned, that is the ability to drill directionally, in some predictable direction other than down. This is achieved by lowering a cast-iron wedge or whipstock down the hole so that the drill bit is diverted when it encounters the slanting hard metal surface. This has great importance in drilling at sea where rigs and platforms are so expensive. Using the technique of directional drilling it is possible to drill many wells starting only 10 feet apart from one platform.

The well bottoms may nevertheless end up perhaps two miles from each other if they are sufficiently deep.

What are the problems in transferring this land drilling technology to the sea?

Very understandably the early underwater drilling techniques did their best to ignore the sea altogether. A sort of barge with hollow legs was constructed, it was towed out to the right spot and flooded to sink to the bottom, piles were then driven into the sea bed through the hollow legs or templates around the edges of the barge. Once these were deep in and secure a platform was built above the level of high tide by welding a flat deck to the protruding tops of the piles, for structural reasons it was necessary to weld the piles to the hollow legs of the template. Derricks and draw-works, etc. then go on top of the flat deck. It sounds laborious but in reality it is a much cheaper way of drilling at sea than any of the others. The driller behaves as if his own deck were dry land and runs his casing from it. The sea is largely ignored.

If the rig has to be moved, a diver has to go down and cut through the piling on the sea bed to free the template. It is not, therefore, very suitable for exploration or for work in very deep water.

Jack-up platforms were developed from the template principle. In these instead of piles the vessel has legs which it racks down until they sit on the sea bottom. The rest of the structure then racks itself up its own legs until it is clear of the waves. This type of rig proved to be popular in the North Sea. Four legs are common but some French designs use five. Number of legs is no indication of safety, *Sea Gem* which was lost in the North Sea had 10 legs.

So long as the rig connects to the sea bed there are no fundamental problems with drilling techniques. If the rig wishes to go away and leave a dry hole, all that is necessary is to fill the hole with cement to sea bed level, cut off the casing running from the sea bed to the surface and leave it at that.

The problems come when the rig is a floating vessel or a semi-submersible device; in these cases there is no rigid mechanical connection between the drilling derrick and the hole in the sea bed. Some idea of the problems can be gained by a brief look at the first attempt to operate a proper drilling ship.

Early experimental work was done in the United States with an ex-Navy patrol craft called *Submarex* in 1953. The CUSS group of oil companies, Continental, Union, Shell, and Superior – which sponsored the work, went on to convert a 268-foot long, 48-foot wide Navy freight barge. The eventual result, launched in 1957, cost 2·7 million dollars, and was named, somewhat unimaginatively, *Cuss I*.

The novel item which was the first piece of real ocean technology was the 'landing base'. This was a steel structure in the shape of a hexagon which had hanging from beneath it the first lengths of well casing. On top of it were blow-out preventers and guide cables for helping the drill string in and out during round trips.

The arrangement to begin drilling **was** as follows: at the derrick head was the swivel block holding up the drill string, then the drilling table and the kelly. The drill string then passed down through the landing base, through blow-out preventers, through 200 feet of casing and ended in the drill bit. The drill string was lowered and where it touched the sea bed drilling began. Once the hole was deep enough the landing stage with its casing was lowered down the drill string to the sea bed. Hopefully the casing slid in the hole and was cemented in by pumping cement down the drill string. The aptly named landing base was then secure on the sea bed and so were the blow-out preventers.

Another pipe came up from the sea bed nearly to the surface; this conductor pipe or riser pipe was responsible for conducting mud from the space between the casing and the drill string inside it back up to the surface. It was sealed at its upper end by rubber packing round the drill string and the returning mud was led off to the side and up to the drill ship by flexible pipe.

The ship itself was held in position by a vast array of buoys and anchors frequently referred to as a cat's cradle. It sounded something of a lashup but it did work and with it came the prospects of drilling for oil and gas anywhere on the continental shelves. Immediately an area at least half the size of Africa was opened up to exploitation.

So by the late fifties the array of gear either on the sea bed or nearby was likely to include large steel valves or blow-out preventers, steel templates with wires and shackles for guiding in drilling strings, various collars and seals to which flexible mud lines could be connected, undersea pipe fittings to conduct oil from producing wells to the surface, and 'Christmas trees', the characteristic arrays of valves at gas and oil heads which control and meter the flow. Further in the future were underwater pipe lines and even underwater storage tanks.

There was a great likelihood of all this gear getting entangled with the cat's cradle of anchor wires and cables or just going wrong, so the demand for diving services to sort out the anticipated difficulties would rise steeply. The money to pay for the development of diving technology was likely to be considerable because the cost of floating a rig was likely to be at least 5,000 pounds a day. Anything to expedite its smooth operation was likely to be paid for handsomely. There was also likely to be a boost to remote-control devices and underwater T.V. cameras for inspection. In America it was also guessed that there would be a need for inspection submarines, underwater jeeps that would buzz about with the engineers to take a look at the installations.

It was into this environment that project Mohole was born. It is interesting to touch on this briefly, partly because it's a nice example of how a big science gambit can misfire, and partly because it highlights some of the problems of operating over the ocean deeps as opposed to the shallow continental shelves.

To recapitulate, the project was to drill a hole down to the layer in the earth's crust where the speed of earthquake waves in the rock changes rapidly. The significance of such a change in speed is that it implies some sudden change in the density of the rocks. This suggests that below the 'discontinuity' the rocks are different either in composition or physical form. Scientists have long referred to this deeper layer as the mantle and they wanted to know what it was made of.

The project originated in a body called The American Miscellaneous Society or AMSOC, a sort of laboriously humorous American scientists' dining club. According to one of the scientists involved, a 1957 review of possible projects before the National Science Foundation had disclosed that there were none that 'would arouse the imagination of the public and which would attract more young men into our science'. The science in question was earth science. One of the members of AMSOC, Walter Munk, suggested that boring a hole down to the mantle would fill the bill. The suggestion was made over breakfast to other members of AMSOC gathered at La Jolla, home of the Scripps Institution of Oceanography. By September 1957 the suggestion had become a resolution of the international Union of Geodesy and Geophysics. The resolution said specifically that a feasibility study should be undertaken.

American enthusiasm for the project was fanned by the fact that the Russians, having scored a bull with the orbiting of Sputnik I, had said that they too were thinking of boring down to the discontinuity (the Moho

View south-westwards of the San Andreas fault, parts of which run through Los Angeles and San Francisco. Displacement of streams and other features is typically several hundred feet. Dark dots in the foreground are tumbleweed halted by a fence. (*Photo: U.S. Geological Survey.*)

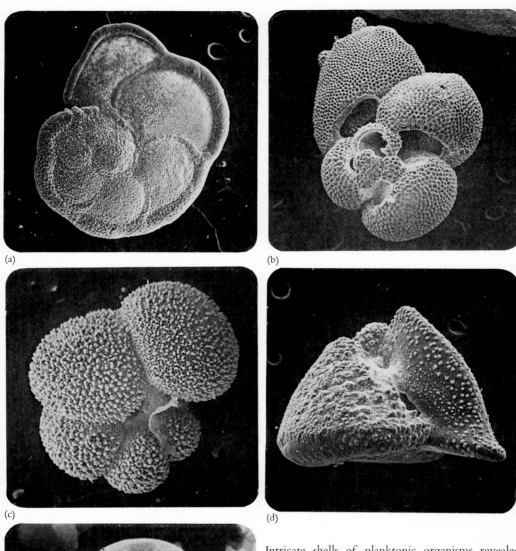

(a)

(b)

(c)

(d)

(e)

Intricate shells of planktonic organisms revealed by the electron microscope, give evidence of the date of sediments and of ancient ocean conditions: (a) to (d) are shells of plankton animals of the Pleistocene period which began two million years ago, (e) is a coccosphere (a plant skeleton) dating from the Eocene period around sixty million years ago. Invisible to the naked eye, it is here magnified 6,000 times. (*Photos: School of Environmental Sciences, University of East Anglia.*)

(a) Globorotalia menardii. Mag. ×64.
(b) Globigerinoides sacculifera. Mag. ×75.
(c) Globigerinella siphorifera. Mag. ×85.
(d) Globorotalia truncatatulinoides. Mag. ×85.
(e) Coccosphere. Mag. ×4,200.

Crude oil underwater storage tank being towed into position for the Dubai Petroleum Company.

Komatsu, a Japanese bulldozer maker, has decided to go underwater. This machine is electrically powered and works down to 60 metres.

One of oceanography's 'jeeps'? A one-man submarine from Vocaline Air Sea Technology, Maine, U.S.A.

after Mohorovičić who discovered it) and that their drilling technology was pretty good. Drilling the hole therefore meant competing with Russians, pushing back the boundaries of technology (since it would have to be a very deep hole) and possibly making great contributions to earth sciences. Geologists were in favour because it might be successful in exploring deep and therefore old sediments on the sea bed and geophysicists were in favour because they wanted to know what the mantle was made of. The reason for choosing the sea as the site was that the earth's crust is much thinner under the ocean so less solid rock would have to be traversed.

The first feasibility study, which itself cost 115,000 dollars, suggested that five million dollars would be enough for what had then come to be called the Mohole.

From the point of view of ocean drilling technology the problem was of a different order of magnitude compared to the offshore continental shelf drilling then being undertaken. Instead of drilling in 300 feet of water, it was proposed to drill in 10,000 feet. For example at one site proposed, Clipperton Island in the Pacific, the depth of water was 10,200 feet, and the Moho discontinuity lay 18,000 feet farther down. Drilling had already penetrated through more rock than that on land but that was not the problem. The major unknown factor was that of how to keep the rig or ship over the hole and how to run the drill string in and out of it.

Anchoring the rig or ship in 10,000 feet of water was soon recognised as more or less impossible and Willard Bascom, then Director of the Mohole Project for the U.S. National Academy of Sciences, decided on the idea of dynamic positioning. The basic idea was to ring the ship with deep moored buoys and maintain its position within the ring by keeping the ship constantly under power. Since ships usually only have propellers at the stern it was decided to add four large outboard diesels to the experimental drilling ship, *Cuss I* which had been chosen for the early trials. These enabled the ship to move sideways.

In the Spring of 1961 five tests holes were drilled in 3,000 feet of water off San Diego and dynamic positioning proved itself. The other matter, that of getting the drill pipe in and out in order to change bits, was not so easy. In fact it may have only just been satisfactorily solved. Current (1970) deep ocean drilling proceeds as far as one expensive drill bit will go; when it is worn out that is the end of drilling.

Indeed even dynamic positioning was not achieved without considerable technical sophistication. One obvious problem occurs when the ship rolls in a wave. Clearly the drill pipe sticking out of the bottom bends. Bascom's early sums showed that when drill pipe is under great tension, as it is when drilling deep holes, a deflection of less than one degree would cause it to kink and snap. His colleague, Edward Horton, devised a trumpet-shaped guide shoe to spread the flexion. Higher strength steel had to be selected too. The same problem occurred at the bottom of the sea where the drilling string entered the sea bed. This time bending could be the result of the ship moving slightly out of position, or a deep ocean current or another effect, the 'Magnus' force, resulting from the rotation of the pipe. The solution was to thread the drill string through a tapering pipe shaped like a hollow fishing rod whose thick end entered the sea floor and functioned as normal casing. It sounds easy enough to say but it required much mathematical work to get just the right taper. Much mechanical ingenuity was required to develop the 'bumper sub', a device used to absorb the vertical heaving motion of the ship so that the drill bit itself did not rise up and down in the hole. Other tricks were required to release the casing and the tapered fishing rod casing from the drill string so that the drilling could proceed normally once the whole assembly was in place on the sea floor.

These problems were difficult enough but they were small compared to the problem of pulling the drill string out to change the bit and then finding the hole again. One way would be for some super submarine to seize the end and steer it the right way but there was no submarine with manipulators that could get to 10,000 feet. Even if there had been it might not have had sufficient power to push 10,000 feet of steel drill string back to its hole. Indeed there was no reliable means of finding any spot on the ocean bed and marking it so that it could be found again.

Just before he left the Mohole project (in 1962) Bascom wrote a book called simply enough, *A Hole in the Bottom of the Sea*. It was a fascinating and topical book which contained a lucid account of the sort of innovations which were likely to be required for the Mohole. Incidentally it was Bascom who coined the name Mohole and he must have regretted it later for, as the costs began to mount, it was easily changed by hostile politicians to, slowhole, nohole, rathole, taxhole . . . etc. Bascom pointed out that a riser pipe would be needed not only to guide the drill string on round trips but to allow a closed circuit of drilling mud to be set up. The pipe would have to rise from the ocean bed nearly to the surface 10,000 feet up and stand by itself. It would have to withstand high pressures of drilling mud and support itself in tension when being lowered into place. One of Bascom's ideas was that this paragon of a pipe should be made of extended aluminium with elongated side chambers in the walls. If these chambers were filled with gasoline, the riser pipe could be made continuously buoyant and would need less heroic means of support at the top. Another novel idea was nylon pipe deployed in the same way. 'The pipe', Bascom observed, 'will have to be firmly anchored to the hard layers within the bottom. If the upward pull of the buoys, deliberately kept large to hold the pipe vertical, should tear the pipe loose from the bottom, the energy of the buoyancy will drive the top of the riser pipe through the well of the drilling ship with', he need hardly have added, 'disastrous results'! To design and install a suitable pipe would, thought Bascom, be a substantial accomplishment.

It was not the only substantial element in the project; the drill ship or rig had to be pretty special too. It had to stay on station, exactly on station, for over a year, it had to keep 100 men happy and fed, and it had to carry a vast draw-works which General Motors was going to provide free. In addition, it had to have a large well cut straight through its middle. Bascom had his eye on a 500-foot long ex-U.S. Navy floating dry dock capable of lifting a destroyer out of the water, that would have needed 'remodelling'.

It was a magnificent scenario; drilling technology would have more 'forced draught' research put into it than it had ever had before. The palaeontologists would get their cores, so would the geophysicists and even the cosmologists. America would get her prestige too, so it was decided that some big company would get the contract actually to do it. The National Science Foundation selected a notable name in the oil business, Brown and Root of Houston, Texas.

Brown and Root decided on a huge platform riding on submerged hulls for stability. The first bill had been 47 million dollars but it soon went up to 127 million dollars although only 20 million was ever spent. By 1966 the project was finally ditched by Congress and entered the history books as a fine (or depressing) example of the interplay of politics and science. Actually science had almost disappeared anyway because the palaeontologists and stratigraphers decided they wanted a lot of shallow holes rather than one deep one, and the geophysicists and petrologists

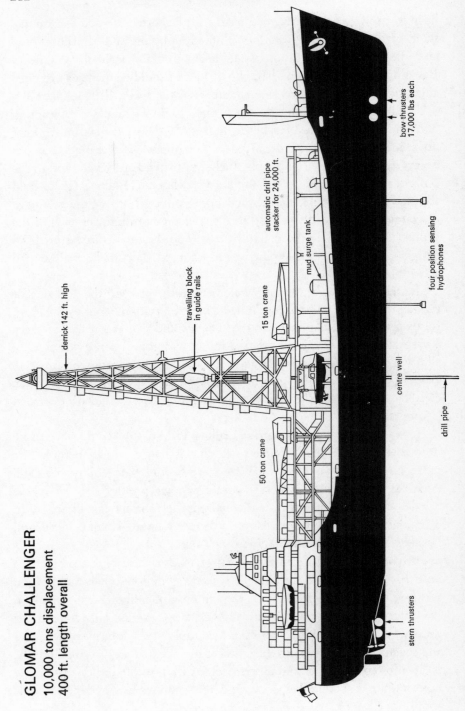

GLOMAR CHALLENGER
10,000 tons displacement
400 ft. length overall

derrick 142 ft. high

travelling block
in guide rails

automatic drill pipe
stacker for 24,000 ft.

mud surge tank

15 ton crane

50 ton crane

bow thrusters
17,000 lbs each

four position sensing
hydrophones

centre well

drill pipe

stern thrusters

were becoming increasingly diverted by the theory of sea floor spreading. As the scientific content of the project was eroded, the political smoke grew thicker. It was attacked because it was going to be a Texan hole drilled at everyone else's expense. Brown and Root was attacked because relatives of the company's president were alleged to have made political contributions of 23,000 dollars and so on. Another reason lay in the fact that the Russians did not really take up the challenge and the moon race took pride of place in international competition.

Before the Mohole project was dead a less ambitious affair which came to be known as 'Joides' was launched by the United States. Joides stands for Joint Oceanographic Institutions for Deep Earth Sampling. The ambition of this project was to drill sedimentary cores in the ocean floor to sample the superficial deposits at many places around the globe. The project turns on the capabilities of the remarkable ship called *Glomar Challenger*. She was built and is now operated by Global Marine for Joides. Her job is to drill holes up to 2,500 feet deep in the sea bed at depths of water up to 20,000 feet. She is less ambitious than the platform envisaged by Brown and Root, which was so big it was dubbed the Texas acre. Nevertheless, the *Glomar Challenger* displaces 10,000 tons. She has sideways thrusters at the bow and stern, amidship is a derrick, and 23,000 feet of drill pipe is ranged on her deck. She navigates by satellite and maintains station above drill holes by placing sonar beacons on the sea bed. The signals from these are picked up by hydrophones and fed to a computer which automatically regulates the ship's position with reference to the beacons.

Drilling is performed without riser pipe or casing. The drill head is simply inserted into the sea bed and lowered until it is dull. The usual practice is to use a coring bit which cuts an annular hole. The cylinder of rock left by the central hole of the bit is periodically retrieved by lowering a wire line down inside the drill pipe. Drilling fluid is usually sea water which escapes into the sea from around the drill pipe on the sea bed or 'mud line'. Once the drill pipe was withdrawn from the hole that was the end of it, there was no way of re-entering it. In 1970, however, a new technique was tried out. A tiny sonar set was pushed through the drill bit after it had been changed and it searched for a sonar reflector placed over the hole on the first drilling run. The string was then manoeuvred into the funnel-shaped reflector by water jets.

Glomar Challenger's second hole is already famous. It was drilled in three days in the middle of August 1968 into the Sigsbee Knolls which lie in 11,000 feet of water in the Gulf of Mexico. The Knolls were thought to

GLOMAR CHALLENGER'S RE-ENTRY SYSTEM

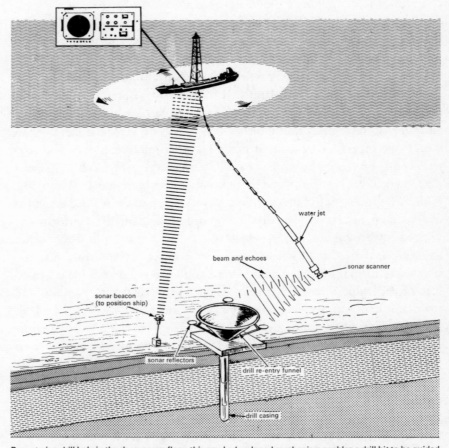

Re-entering drill hole in the deep ocean floor, this newly developed mechanism enables a drill bit to be guided into a previously drilled hole. The re-entry funnel is attached to the drill string when it is first lowered to the bottom, and remains on the ocean floor when the string is withdrawn. At the time re-entry is attempted the drill string is relowered with a sonar scanner on the bit assembly that emits sound signals. These signals are echoed back from three reflectors spaced around the funnel. Position information is relayed to the ship and the water jet is used to steer the bit directly over the funnel.

be salt domes rather like the ones where, on land or on the continental shelf, oil is sometimes found. Had *Glomar Challenger* drilled through into oil-bearing strata it might have been disastrous since she had no means of preventing blow-out. Therefore when oil and gas saturated limestone cap rock was recovered only 450 feet below the sea bed it was decided to stop drilling.

If there is oil or gas on the deep ocean floor and it proves to be wide-spread, then the consequences for the oil industry and for deep ocean and deep drilling technology are great indeed.

But so far the Mohole and the Joides projects can be regarded as being tangential to the main stream of development of drilling. Even if *Glomar Challenger* can re-enter a hole, there is still the problem of riser pipe to be solved.

Commercial drilling will not make any great leaps forward, instead it inches out into deeper water step by step. The favoured means is not so much the drilling ship as the semi-submersible rig. One of the latest, launched in January 1970 by the Japanese firm Sasebo for Transworld Drilling, has four massive 160 foot high legs 270 feet apart. Its feet are large pontoons which can be flooded up so that they reach equilibrium 100 feet below the waves, 'floating' undisturbed whatever the weather on top. The platform of the rig will then be 30 feet up above sea level. Casing can be run to depths of 20,000 feet below the sea bed. In the stormy North Sea off Western Europe semi-submersibles have been widely used either in the semi-submerged mode or actually sitting on the bottom.

Clearly, however, the deeper the water the greater the problems of doing things at the well head. Fixing permanent surface platforms to the sea bed 600 feet down is getting too difficult and expensive and increasingly the trend is to put the well head on the sea bed. Doing this by remote control, peering through closed-circuit T.V. cameras is getting more and more difficult. The very human urge is to send someone down to fix it. Getting a man down there to do something is a whole new area of technology and it forms the subject of the next chapter. Unlike the nineteenth-century world of drilling with steel, grease, sweat and heavy manual labour in the open air, the twentieth-century world of high pressures and environmental control is rather more scientific and far less predictable.

I2

Send a Man Down

IT IS DIFFICULT TO TALK ABOUT DIVERS WITHOUT FEELING THAT ONE IS ABOUT to slander those one knows and admires. Divers are the product of natural selection since even in the world's navies they are usually volunteers. Who but a very special sort of chap would volunteer to descend into the dank, cold, dark and desperately lonely depths to do the sort of things that usually need doing? In shallow water the jobs are things like freeing propellers from ropes and heavy lengths of weed, scrabbling junk out of the hinges of lock gates and sluice gate seatings, welding a new piece onto a ship's bottom, or clearing out a water intake. Divers are nearly always called upon when something goes wrong or gets fouled up. The boss of one British diving firm says that, frankly, they have to be brave underwater navvies. Unlike their opposite numbers, test pilots and astronauts, who have the good fortune to live at low pressure, divers rarely have to perform complex tasks. In deep diving it is sufficient if somebody can go down, get an idea of the shambles and either stick some dynamite in the right place or lash a lifting wire round something suitable and retire. Tasks like changing valves or inspection can often be done better by remote control or underwater T.V. Jobs that might be routine on the surface become very difficult indeed in cold muddy darkness, in strong currents, when the diver is greatly befuddled by the very fact of breathing and existing in a high pressure environment.

At the Oceanology '69 conference at Brighton, England, a French expert Dr. P. Fructus from the *Compagnie Maritime d'Expertises* (COMEX) told his audience of experiments with human volunteers on very deep dives over 800 feet. It can readily be seen that very deep diving requires a special enthusiasm. On May 14th, 1968, Messrs. Fructus and Jullien dived to 1,000 feet in 85 minutes and stayed there 20 minutes. They took 91 hours to decompress back to atmospheric pressure. On the way down they complained of narcosis, confusion and trembling. Aches and cracking of the joints also featured, the Americanism 'no joint juice' was said to describe the feeling rather aptly. On the way back general tiredness and multiple cramps upset their sleep. The 'bends' were a recurrent frisson, the lecturer reported that he could feel bubbles passing through both knees and the left ankle.

During the next dives they feared that trembling might be an early warning of convulsions (having extrapolated the results of experiments performed on animals) therefore precautions were taken, sharp items inside the diving chamber were padded. In a few concluding remarks, Dr. Fructus mentioned that bone necrosis had so far only been observed in caisson workers and may not attack divers at all.

This kind of experience gave rise to the notion of the so-called 'helium barrier'. It was believed that 1,200 feet represented the greatest pressure at which oxygen–helium mixtures could be breathed. The greatest pressure for safe breathing of oxygen–nitrogen mixtures or ordinary air was much less, around 200 feet. The Royal Navy didn't believe in the helium barrier and in January 1970 two volunteers went to a depth equivalent of 1,500 feet breathing oxygen and helium at 45 atmospheres pressure. *Nature* reported 'there is no sign yet of any physiological limit for diving and at least a further 300 feet of the sea are opened up to explorers who take the necessary precautions'.

The need for these physiological heroics is occasioned as much by a spirit of scientific enquiry as by the pressure of commercial demand. So far the latter has encouraged commercial diving only to 500 to 600 feet and such depths are the exception rather than the rule. The protagonists of diving are undeterred by the difficulties and point out that although remote-controlled manipulators and T.V. are useful if the water is clear, only divers' sense of touch will do when conditions are bad. But other difficulties are legion: for example divers when breathing oxy-helium mixtures talk like Donald Duck; special electronic receivers have had to be developed to unscramble their speech for surface listeners. Some divers have claimed that after a few days living at high pressure they can understand each other but the usual communication is by hand signals, grunts, exaggerated nods, and by slowly writing messages on slates which are then deliberately presented in front of the other man's face mask.

It is all rather primitive and unsatisfactory but diving services have been commercially far more important than all the other techniques of doing things underwater. No matter how unintelligible, clumsy, befuddled and cold a trained diver may become, he is generally more efficient and economic than many of the machines designed to do the equivalent jobs, but he must have the right tools and the right support and the water must not be too deep.

There is also a romantic symbolism about diving. Going to the deeps in an air-conditioned submarine is not really living in the sea. Cousteau, in whom the streak of romance is marked, was the first to establish a 'house under the sea' called Conshelf, where men lived a so-called normal aquatic life at the pressure of the sea, passing in and out of the water into a submerged bell. Other people have taken this idea further since the invention in 1965 of American General Electric's selectively permeable membrane. This membrane allows oxygen normally dissolved in sea water to pass as a gas across it to regions where oxygen pressure is lower. Therefore in theory man can be like the fishes and breath below the surface with artificial gills. The application has already been patented in the United States. Hamsters have lived happily under water for two weeks without external air supplies and it should be possible for a man to do the same, theoretically, inside something as small as a four foot cube of the membrane. It might even be possible for a man to flood his lungs and breathe water.

According to N. C. Fleming of the National Institute of Oceanography the time lag between a laboratory experiment with men and an experimental sea dive is only two or three years. A similar period of time elapses before commercial applications are possible. On that principle if the Royal Navy's 10 hours at 1,500 feet in 1970 is anything to go by, 1978 should see divers working on oil wells 1,000 feet deep or more, always supposing that the oil industry finds it economic or politic to go so deep.

The oil industry has backed all the horses, it is the only way of being sure of winning the race. As long ago as 1960 Shell had a robot with a socket wrench and an underwater T.V. eye going down to attend to well heads. In 1964 Shell signed a research agreement with one of the big names

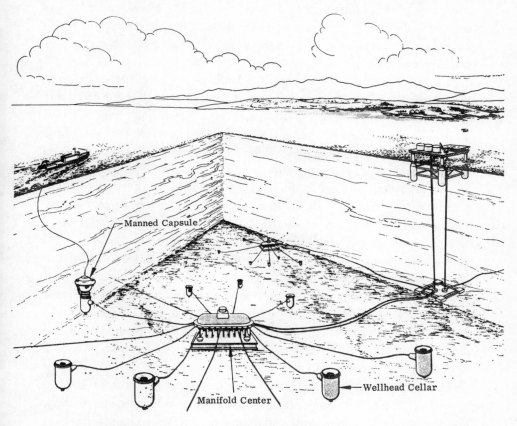

LOCKHEED'S OIL SYSTEM: Surface ship (*left*) deploys manned capsule to descend through ocean and 'mate' with wellhead cellar mounted on undersea well. Other cellars are shown feeding their fluids through lines into a manifold centre, where liquids would be mingled and passed through other lines to surface separation facility (*right*).

in diving, Dr. Hannes Keller, who had made an open sea dive to 1,000 feet in 1962 and survived. The research tool was an advanced diving bell called Capshell. Most diving bells consist of a dry pressure vessel which is lowered to operating depth and pumped up until pressure inside is equal to that outside. Divers then open the door and swim out to work. When they return they seal the capsule to maintain the pressure, the capsule is then winched to the surface and mated with another larger and more comfortable pressure vessel on deck in which the divers can pass the hours or days of slow decompression in comfort. Shell's device consisted of a doughnut-shaped ring of pressure vessels wrapped round a conical skirt. The whole device is lowered over a well head and air is pumped in until the water level is pushed down to expose the job to be done. The divers emerge and work in the dry, breathing oxy-helium mixture from hoses. One chamber in the torus stays at atmospheric pressure so that supervisors can view and control the proceedings.

So far there are no ways round the fundamental problems of living at high pressures. It is now believed that the old 'rapture of the deep' is due to nitrogen dissolving in body tissues and producing symptoms of narcosis. Even helium which has a low solubility, may do the same if the pressure is high enough. Helium which is one seventh as dense as nitrogen has the virtue of reducing the effort required to breathe at high pressure. Its vice is that being a good conductor of heat it causes divers to get cold more quickly. Another drawback is that it is so expensive that diving companies ruefully claim it costs a dollar a breath. Nevertheless, the use of oxy-helium has put diving in business throughout most of the continental shelf areas and made worth while a whole satellite technology designed to help divers be more efficient, but more of that later.

Several American companies have decided on a quite different answer to the problem of trouble-shooting on the sea bed. The advanced technology companies have shown marked reluctance to get men wet and really live in the sea. For most of them the aim has been to provide a 'shirt-sleeve' environment in which men work most efficiently.

Lockheed, trying hard to diversify out of aerospace, got itself into the oil and gas business by establishing Offshore Petroleum Systems in 1969. The aim was to let a man work on ocean floor oil wells as comfortably as he does on land.

When an oil or gas field has been discovered it is necessary to drill producer wells all over it; how many depends on the production required. The flow from the wells is piped to a central control and mixing array. Each well has a so-called Christmas tree of valves which can send the gas in various directions; often gas or oil comes up a separate tube inside the casing and the space between the tube and the case will then be connected to another valve. Sometimes wells need to be reworked because of blockages or because a blow-out preventer or a storm choke deep in the well is faulty. Whatever the cause a producer well needs to be maintained.

Whether the well heads are on the surface of the sea or on the sea bed, running them will cost money and the deeper they are the more they cost, either because a taller production rig is needed or because deep diving is more expensive, or both.

Lockheed's answer is to fit a large steel cellar on to each well head on the sea bed. Each cellar is 21 feet high and 8·5 feet wide and is full of air at one atmosphere pressure. It is slid into place down the wires from the drill ship or rig to the landing base which, it will be recalled, enables the drill string to be run in and out of the hole. The cellar fits on to the well head. Next on the scene comes a utility chamber or capsule attached by an umbilicus to a ship on the surface. The capsule has little propellers and swims down to the cellar. It docks with the cellar using technology developed by Lockheed for the Deep Submergence Rescue Vehicle, a Navy funded vessel for rescuing stranded submariners. Workmen then climb down into the cellar and fit the necessary valves. Pipe lines are fitted, again without anyone leaving atmospheric pressure or getting wet. The pipe is winched down from the pipe line barge and fits into a special gland. The design of the connector system, says Lockheed, accommodates ejection and replacement of a damaged flow line bundle.

Lockheed's engaging little diagram shows a sea bed sprouting with cellars looking like bottle-shaped flowers all connected by pipes to a master cellar or manifold with mixing facilities. Like errant bees the manned capsules are swimming down trailing their air and power lines from surface ships. Lockheed demonstrated the technique in 1970 in the Pacific. It is designed for operation at 1,200 feet or more.

Another system for which great economies are claimed is that of Deep Oil Technology Inc., California. Deep Oil is 80 per cent Fluor Corporation and 20 per cent Ocean Science and Engineering (O.S.E.). The latter company is itself partly owned by Fluor and all sorts of other large companies who want a look in on deep ocean technology; they include Anglo-American, Amerada and the Aluminium Co. of America among others. The biggest single shareholder and leading light is Willard Bascom who founded the company when he left the Mohole project. O.S.E. began building Deep Oil's sea bed system at Long Beach in 1969. Most of the finance comes from Fluor which until 1967 had very little to do with the sea. It was a chemical engineering company and, like the aerospace companies, it got interested in oceanography in the general rise of interest. Unlike Lockheed, however, Fluor got its feet wet by acquisition, buying

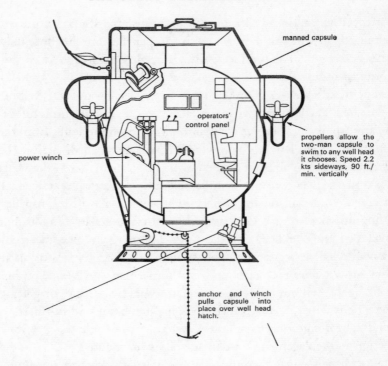

manned capsule

operators' control panel

propellers allow the two-man capsule to swim to any well head it chooses. Speed 2.2 kts sideways, 90 ft./ min. vertically

power winch

anchor and winch pulls capsule into place over well head hatch.

LOCKHEED'S DEEP WATER OIL PRODUCTION SYSTEM

for depths down to 1,200 ft.

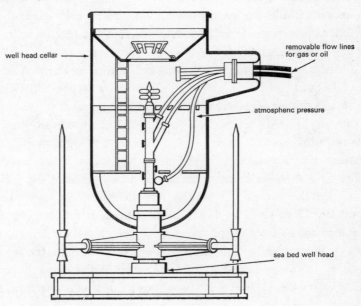

well head cellar

removable flow lines for gas or oil

atmospheric pressure

sea bed well head

into or buying up companies that were already in operation. It concentrated on offshore drilling companies and the Deep Oil enterprise nicely combines pressure vessels and drilling technology.

The system is designed for production wells where the practice is to drill a number of wells into the same field. Using directional drilling these can be drilled close together. The idea is to put a template on the sea bed with say 10 cylindrical well heads in it. A submerged work chamber moves past these on guide rails and extends from itself tools suitable for working on the well head valves and preventers. Inside is the desired shirt-sleeve environment and the operators illuminate their work with lights and view it with T.V. if necessary. The work is necessarily simple, and defective equipment will be replaced rather than repaired. The work chamber is about 11 feet long and is just less than six feet in diameter. If anything goes wrong it can pop up to the surface under its own buoyancy. Normally it propels itself down from surface and back up again by two propellers. Power supplies come from the surface by cable and enable the tools to be manoeuvred with great force; up to 10,000 pounds of torque can be exerted and weights of several tons can be lifted.

This Deep Oil system, now completed, can work in conjunction with a special kind of semi-submersible surface rig called a tension leg platform. Between them they allow wells to be drilled at depths down to 1,500 feet for the same cost as conventional wells at 400 feet.

Lockheed's and Deep Oil's systems are two amongst several and any comparison of their merits must wait for operational experience. But they compete off the drawing boards as aircraft do, one system is less expensive to install but more expensive to operate, another is very flexible and yet another suitable only for production field development and so on. Deep Oil Technology's system will sell for around half a million dollars.

The Japanese have the most exotic plan of all which makes the rest of the underwater oil and gas drilling systems look like small beer. All the systems so far developed still depend on a surface drilling ship or rig with its derrick and draw-works. No-one ever suggested putting all that underwater with the necessary thousands of feet of drill pipe as well; no-one that is, until the Japanese began to consider drilling in the Sea of Japan.

Japan, now the third biggest industrial nation in the world, has no natural resources of her own except a certain amount of coal. She imports an enormous amount of oil and in Japan importing is still an economic sin

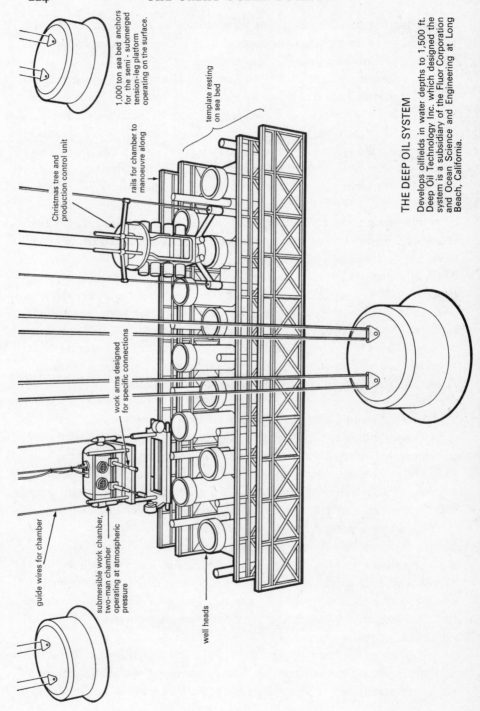

1,000 ton sea bed anchors for the semi - submerged tension-leg platform operating on the surface.

template resting on sea bed

Christmas tree and production control unit

rails for chamber to manoeuvre along

work arms designed for specific connections

guide wires for chamber

submersible work chamber, two-man chamber operating at atmospheric pressure

well heads

THE DEEP OIL SYSTEM

Develops oilfields in water depths to 1,500 ft. Deep Oil Technology Inc. which designed the system is a subsidiary of the Fluor Corporation and Ocean Science and Engineering at Long Beach, California.

She does have access to a large continental shelf both between Japan and Korea and into the East China Sea towards Formosa (Taiwan). Recent surveys of the latter area seem to show that it could contain very large reserves of oil and gas. Japan's planners, however, have been looking nearer home at the Sea of Japan. The Japanese maintain that it is even stormier than the North Sea and that operating with surface rigs will be greatly hazarded by bad weather. This means that it will be much more expensive. If a surface rig is put underwater it seems at first to pose colossal problems in pressure vessel design. A standard drill pipe length is 30 feet and when making a round trip it is very desirable that two or three lengths at a time should be handled. Therefore a very large and expensive pressure vessel would be needed. However, drilling is very much a repetitive job; if ever a job cried out to be automated it is round-tripping 10,000 feet of drill string 100 feet at a time. The Japanese have therefore funded a radical investigation into the problem of the remote control and automation of drilling and the resultant possibilities of putting the equipment on the ocean bed at ambient pressures. Pressure vessels containing shirt-sleeved engineers at atmospheric pressure might be reducible to an economic size and could fit inside a lighter and larger structure. Needless to say there are many problems, especially the provision of power for the draw-works and such items as mud screens where geologists pick off the rock chippings which have come up with the mud and analyse them to control the progress of the drilling.

Anyone who proposes to have a quiet smile at this approach should remember that the Japanese have a tradition of squeezing the best out of established technology as their railways and large tankers already testify.

Perhaps at this point it is important to introduce a large digression. Having a man down to fix things is not just a likely requirement of drilling rigs and sea bed Christmas trees. The excursion of the oil and gas business into deeper waters has all sorts of ramifications which require engineering and maintenance to be done underwater. Not only is there an extraordinary diversity of hardware in view, but, because it is so expensive, the need to keep it in trim is more and more pressing. It is interesting to look briefly at some of the hardware actively being designed or made. It indicates that man's advance into the sea, even if it were only to be fuelled by the commercial requirements of the oil and gas companies, will be on a far wider front than the rig and well maintenance so far outlined. Many of the structures currently being proposed will not be within the scope of specially designed work chambers with a preplanned and limited series of possible manipulations.

Take for example the consequences of piping oil ashore to await the arrival of a supertanker.

When the oil companies add the cost of laying a pipe, and probably having immediately to repair it because somebody has put an anchor through it, to the cost of building terminal tanks on a distant shore, and then the cost of a jetty to receive the tankers, they ought to look with interest at underwater oil storage. Although this sounds fantastic and laughably expensive it has a great deal to be said for it and several installations are already in existence.

Since oil floats on water a suitable device can be bell-shaped and open at the bottom. Its walls do not have to withstand the enormous pressures so usual in underwater technology. There are nevertheless substantial forces to be dealt with, one proposed tank designed to store 500,000 barrels of oil has to withstand around 15,000 tons of buoyancy thrusting upwards. This size of tank is reportedly the minimum useful size, if it were on shore it would have a diameter of at least 250 feet and would be 60 feet high. A design proposed by the Chicago Bridge and Iron Company is shaped rather like a T.V. tube face downwards, it is 200 feet high but most of the bulk would be 100 feet deep. If supertankers drawing 60-70 feet are to moor at its top to load the oil, most of the bulk would necessarily have to be well submerged. The lowest segment of the structure would be double walls of steel filled with concrete to keep them down.

Such vast structures have to be built on shore like a ship. Indeed they are comparable to a 10,000 or 15,000 ton ship. They are usually built behind dykes and when completed they are floated off by pumping sea water into the construction basin. Once afloat the structure is towed into a deeper portion; the water level is then lowered to sea level, the dykes are breached and the tank is towed out to sea.

When at the desired spot it has to be carefully sunk into place. Once it is on the sea bed piles are driven round the skirt to help it withstand the stresses to come. The strength of the structure is given a severe test during submergence and in Chicago Bridge's design there is sufficient excess strength once the tank is in place to build a production platform on the top. Indeed it could even support a drilling rig so that it would be possible to drill for oil through its base. It is not for exploration clearly but it could be used to develop a small field once it had been found and proved.

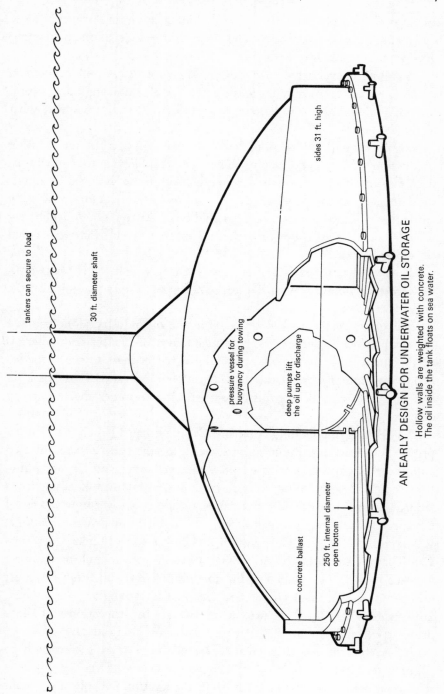

tankers can secure to load

30 ft. diameter shaft

sides 31 ft. high

pressure vessel for
buoyancy during towing

deep pumps lift
the oil up for discharge

concrete ballast

250 ft. internal diameter
open bottom

AN EARLY DESIGN FOR UNDERWATER OIL STORAGE

Hollow walls are weighted with concrete.
The oil inside the tank floats on sea water.

Such structures are in for a rough life since passing waves impose fluctuating pressures on them. The possibility of one of them bursting does not bear thinking about.

All this, however heroic it sounds, is just shallow water stuff, occasioned as much by the needs of tankers as by the fact of producing oil at depth. Suppose one has an oil well 1,000 feet deep, well out from shore, what then?

Ocean technology is not daunted; one solution is the floating production and storage facility. This may be likened to a medium-sized tanker, say, an 80,000 tonner floating with its bows in the air and its stern 400 feet farther down. The freeboard of such a structure, that is the amount standing above the waves, would be about 100 feet. A drilling rig would be cantilevered out over the side so that production wells could be 'worked-over'.

The French company ELF has proposed a rig which waves backwards, forwards and sideways with the sea, gyrating on a universal joint on the sea bed.

It is quite clear that developments in the oil industry alone provide many things to go wrong or to be maintained on the sea bed. There is ample impetus for the technology required to get men down on the sea bed to do a job of work without taking into consideration futuristic ideas like underwater bases and missile silos that come under the heading of defence.

The British shipbuilding company Cammell Laird has embarked on a project which seems to unite or at least bridge the gap between the two approaches looked at in this chapter, the wet diver and isolated shirt-sleeve engineer with remote control. Bridging gaps is but a short step from falling between two stools and it is not possible to say yet how successful the approach will be. Cammell Laird, with the help of Britain's National Research Development Corporation, has built a sort of underwater Sno-Cat. That is to say, that just as a Sno-Cat was a vehicle built for crossing ice caps that people could live in, do things from, and leave for short excursions, so Cammell Laird's Seal Beaver can move over the sea bed, keep its crew safe for days, provide a platform for inspection and allow divers to exit and enter so that they can work close alongside.

It is a bizarre and original vehicle and looks fearsome enough to frighten the fish. Its single eye is a sonar scanner which comes up on a telescopic pipe from just in front of an observation cockpit. Two arms

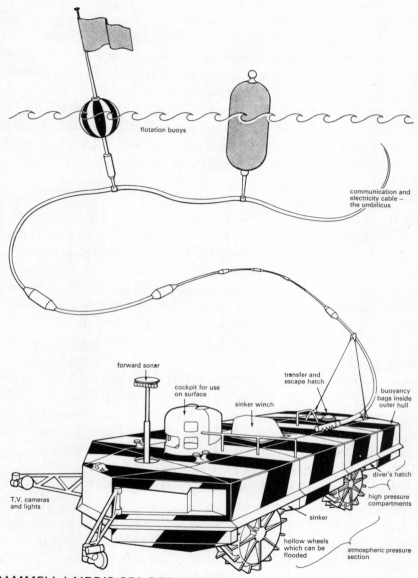

flotation buoys

communication and
electricity cable —
the umbilicus

forward sonar

cockpit for use
on surface

transfer and
escape hatch

buoyancy
bags inside
outer hull

sinker winch

diver's hatch

T.V. cameras
and lights

high pressure
compartments

sinker

hollow wheels
which can be
flooded

atmospheric pressure
section

CAMMELL LAIRD'S SEA BED VEHICLE
maximum speed — 2.5 kts
maximum crew — 10
design depth capacity — 600 ft.

with T.V. cameras or lights protrude from the front, and the umbilical cable comes out of the back. It moves on four large wheels which are really hollow drums. Inside its rectangular body is a pressure vessel divided into two. The forward one is maintained at atmospheric pressure and is for the operating crew and observers. It has most of the vital electrical equipment together with cooking and toilet facilities. The rear compartment also has cooking and toilet facilities, and is designed to function as a decompression chamber for the divers. Farther to the rear is a transfer pressure vessel fitted with top and bottom hatches. The lower hatch gives access to the sea bed whilst the upper one can communicate to a transfer bell lowered from the surface. There is also an access hatch to the forward compartment from outside.

The pressure vessels are inside a rectangular steel frame. The space between the two contains buoyancy bags. The whole vehicle is positively buoyant and will float, it gets to its destination by being towed on the surface with the four wheels retracted inside the rectangular outer shell. Once over the operating site the vehicle submerges by juggling with ballast tanks and sinkers. The proceedings begin by levering a sinker to rest on the sea bed and putting the wheels into their operating positions protruding from the body. Ballast tanks are flooded until the vehicle is just buoyant. Its sinker is now heavier than the remaining buoyancy so it can winch itself down to the sea bed. When it arrives and the wheels touch, it continues to winch on the sinker until the sinker is close up against the body of the vehicle. The sinker's weight is now applied to the sea bed through the four wheels. The reaction between the sea bed and the wheels is then one ton for each wheel. In emergency it is possible to cut the sinker loose, the vehicle should then rise to the surface under its buoyancy, even without blowing the ballast tanks.

The wheels are eight feet in diameter and three feet thick, they are thus very large in volume compared to their loading and make it possible to traverse extremely soft ground. Cammell Laird claims that sediment without any cohesion with a density of only 60 pounds per cubic foot would support the wheels by displacement, they would not sink in more than three feet or so. In this mode the vehicle would be moving like a sort of mud-bound paddle steamer. The makers believe it can cross any submarine terrain except steep rock and freshly deposited estuarine silt.

The vehicle is about 50 feet long, 20 feet across and 16 feet high when standing on its wheels. It weighs about 90,000 pounds and can carry up to 10 men. Its speed is two and a half knots, it will go up a 50 per cent slope and pull between 5,000 and 10,000 pounds on the level depending on the terrain. It navigates by dead reckoning, all done automatically by a gyro-compass and a computer. Everything unbolts so that other modules can be quickly attached for jobs such as trenching or sampling. It can also be broken down and transplanted in standard containers.

What will it do? Cammell Laird intend to hire it out and at first sight it might seem destined for the fate suffered by so many small American submersibles and at least one British one, lying ashore unused because of being too expensive.

Obviously salvage and unforeseen repairs will provide one market, but another may come as rigs move into deeper waters. The legs of drilling rigs, especially permanent production platforms, will be so long that they will need bracing. In shallower depths, major civil engineering projects like the building of offshore islands for bulk cargo storage at Dunkirk or in the Irish Sea will need to give divers the things that ordinary con-struction workers want and get on an ordinary building site—machinery, power on tap, and somewhere to shelter.

All these developments have one thing in common, they have happened for commercial or economic reasons. They owe very little to all the demands of defence, or scientific research, or technological gymnastics of the kind sometimes sponsored by the Americans, the French or the Russians to show what can be done. It is important to separate the cate-gories because in the present mood of the times commercial activity tends to be regarded as palpably honest, whereas activity undertaken for other reasons and requiring taxpayers' support tends to be regarded as more suspect. The proposition that the oceans are in some sense man's last frontier is thus demonstrated and there can be little argument about it. It is clear that the frontier such as it is will go on being pushed back and that the process will compel all manner of adjustments in the attitudes of governments, particularly on the question of who is 'legally' entitled to what.

The big science lobby has favoured the development of technologies and projects which are not strictly commercial but which will nevertheless have considerable bearing on the development of ocean exploration. Sometimes these come under the heading of defence, sometimes not. They form the subject of the next chapter.

13

A Job for Machinery

IN THE SIXTIES THE RESEARCH SUBMARINES BECAME THE SYMBOL OF OCEANO-
GRAPHY. Most of the companies in the United States who had pretensions
to 'ocean capability' acquired one, sometimes more than one. There were
some fairly absurd ceremonies to mark their progress. The Westinghouse
Deep Star went down to 4,132 feet, planted a flag in the bottom and took a
picture of it. Subsequently a sonorous advert appeared in *Time* magazine,
'The *Deep Star* log on May 11th, 1966 recorded, " ... everything is working
perfectly. We will now plant the flag in the mud using the mechanical
arm ...".' Later on Lockheed planted the Stars and Stripes at 8,000 feet.
General Dynamics, North American, General Motors, Grumman,
Reynolds Aluminium, all had their own vessels and other companies got
in on the act by advertising that so-and-so's submarine used their product.

In Britain, the fact that only one was planned, and that was being built
on a shoestring, was a reason to berate the government and industry. Even
Switzerland acquired a modest one and France made the impressive deep-
diving *Archimède*. A research submarine was the jeep of the deep ocean and
one was actually called 'deep jeep'. In the latter part of the sixties, the fact
that most of the little submarines were out of the water and on the stocks
was widely regarded as proof that the expected oceanographic boom had
misfired. However, things were not quite as simple as that. Many had been

built in the competition for the Deep Submergence Rescue Vehicle (D.S.R.V.) project and were never meant to earn a living by being hired to offshore oil companies. Others were built to be test vehicles for oceanographic systems and instruments designed for a market that the U.S. government has not yet funded into existence. Some critics too would assert that because of fundamental technical limitations, the submarine is not the jeep or the Land-Rover of the ocean depths. Nevertheless, if ocean exploration is to be undertaken seriously and comprehensively, it is arguable that it will have to be by some sort of submarine.

It is easy to forget how much of our picture of the ocean bed is based on the point sampling and indirect geophysical exploration or sonar viewing. The pilot of the Vickers *Pisces* research submarine (built in Vancouver) for example, reporting on his experiences in Loch Ness and the Sound of Mull, stresses that what you actually see on the bottom bears little relation to the chart. The smooth contours are merely hopeful lines linking point sources of information. Cruising over the bottom looking through the window reveals that the sea-bedscape has got unsuspected 200 foot cliffs and chasms. The first time the submarine submerged into Loch Ness, she went 100 feet deeper than the deepest sounding on the chart. Observation showed that the bottom was largely peat and the water was the colour of weak tea. There was so little life anywhere in sight that the well-publicised monster must live a very hungry existence indeed. Cruising through the Sound of Mull the sea bed turned out to be strewn with areas of huge boulders. Here again the sea bed is not the soggy plain that, say the operators, everyone had imagined from the previous indirect observations. The supporters of research submarines declare that there is no substitute for getting down and having a look. The trouble is that doing so costs a lot of money. In the case of the Vickers *Pisces* and its attendant, 1,000 pounds a day. This vehicle will go down to 5,000 feet with two men, and is a good example of the medium performance research submarine.

In many ways, the economics of submarines seem to be very like those of aircraft, no-one would use a long-range Boeing 707 for commuting between Santa Catalina Is. and Los Angeles or a Britten-Norman Islander for crossing the Atlantic. Either machine may well be capable of doing it, but the costs are hardly attractive. Using the expensive deep-diving *Pisces* to watch the daily habits of shrimps at 300 feet is not doing a scientifically necessary task the cheapest way. But the market is still so small that specialisation is not worth while, except in the oil industry inspection market.

Tracing back the earliest 'research' submarine probably leads to Cousteau's Diving Saucer *Denise* in the late fifties, though two-man midget submarines were used by the Royal Navy, the Italians, the Germans and the Japanese in World War II. The major impetus in the United States came with the loss of the U.S. submarine *Thresher* in 8,400 feet of water in the North Atlantic during April 1963. There were only one or two vessels capable of search and salvage operations, and neither of them were of American design or manufacture. They were the French *Archimède* and the bathyscaphe *Trieste*. As a direct response to the tragic loss, the U.S. Navy formed the Deep Submergence Systems Review Group (D.S.S.R.G.) to look into the whole question of the Navy's capacity for finding, recognising and recovering deeply submerged objects. The Review Group concluded that the established systems were too dependent on good surface weather; that they wouldn't work under ice nor would they work in really deep waters. A Deep Submergence Systems Project (D.S.S.P.) was established in 1964 at about the same time as the new enthusiasm for oceanography gathered pace. The first solid contract on offer was that for a D.S.R.V. or Deep Submergence Rescue Vehicle. There were hints that by July 1968 there would also be contracts for five more 'tactical' versions of the vehicle. It looked like a solid line of business and the competition to get in was intense.

In order to prove their submarine capabilities, several companies built their own vehicles and the Rolls-Royce of them all was Lockheed's, named *Deep Quest*. Lockheed has a company text to steady corporate nerves 'Look ahead where the horizons are absolutely unlimited'. At the time of writing, Lockheed had progressed further towards the horizon than Congress was willing to pay for, notably with the Jumbo freighter C5A, but there is no doubt that Lockheed made a magnificent effort in its bid to get the lion's share of the technology contracts on the deep ocean horizon. The D.S.R.V. contract called for the vehicle to be able to 'mate' with a rescue hatch in a damaged submarine at heel angles up to 45°. The transfer passage had to be sealed and dewatered to allow the dry transfer of men at depths down to 3,500 feet. The D.S.R.V. was to be light enough to be air freighted and was to be carried to the site by another nuclear submarine to which it would transfer the rescued men whilst still underwater.

Deep Quest was designed to show that Lockheed could make the D.S.R.V. It is made of two pressure spheres of maraging steel. This material has a tensile strength of 180,000 to 210,000 pounds per square inch and is notably superior to the usual submarine hull steels which have tensile strengths of 80,000 pounds per square inch. Even 'ordinary' submarine steels present difficult enough welding problems. With maraging steel, the whole sphere is fabricated first then fitted into a furnace for heat treatment; the process is known as ageing, and involves the precipitation of microscopic particles through the matrix. Thanks to these highly expensive techniques borrowed from rocket motor technology, *Deep Quest*'s pressure hull is notably thin ($\frac{3}{4}''$) and light. The only penetrations apart from the view ports and access hatch are for electrical connections. The whole craft is 40 feet long, weighs about 50 tons, and is, once again, 'all singing, all dancing'. Stepping on board feels like climbing into a space capsule. The place is crammed with instrumentation and controls and the walls of the two seven-foot diameter capsules are almost totally obscured except for the small portholes under foot. The men who handle her are referred to as pilots and it is quite clear that, in terms of judgment and co-ordination, if not in speed of reaction, that is exactly what they need to be. The vessel has two propellers and the usual rudders and planes but it also has bow and stern vertical thrusters and lateral water jets. It has on-board computer facilities that allow it to navigate by reference to sound beacons placed on the bottom. It has doppler sonar to tell its true speed over the ground and a large array of lights and T.V. cameras. It can take geological cores, pick up objects and manipulate them. It even has a diver lockout capability. It has power for 12 hours at four knots and life support for 190 man hours, and it cost Lockheed an absolute fortune.

With it Lockheed won the contract for the D.S.R.V. and would now have a commanding position in the deep ocean exploration industry, if only there was one. The D.S.R.V. itself is as remarkable as *Deep Quest*. It will dive to 5,000 feet beyond the collapse depth of operational military submarines. It carries 24 survivors at a time and can be airlifted to the scene of any disaster. According to a local congressman who spoke at the San Diego launch in January 1970, D.S.R.V.'s tasks of location, mating and rescue 'in many ways exceed the complexity of the lunar landing problem'.

After the airlift, D.S.R.V. gets right to the scene of disaster by riding piggy back on one of the many large Navy nuclear submarines specially fitted with pylons. It is to this submarine that survivors are transferred underwater, or even under the polar ice which is a favourite place for Polaris submarines to shake off their Russian shadowers.

D.S.R.V. is made of HY 140 steel, i.e., steel with a tensile strength of 140,000 pounds per square inch. However, there are many other materials which might be even better for deep submergence; strength is not the only requirement. Ideally one needs the displacement of the hull to be around twice the weight so that positive buoyancy is preserved. Therefore, as in aviation, lighter materials have their attractions. Aluminium was used by Reynolds Metal in making *Aluminaut* (Electric Boat did the actual building) which is quite a large submarine with a depth capability (untested because of insurance problems) of 15,000 feet. Titanium is another metal with high potential, but in 1965 both were rejected by the Navy Special Projects Office on the grounds that both needed careful and continuous attention in the corrosive environment of the sea. Apparently, solid glass is a much more attractive candidate with glass fibres in polyester resin a close competitor. Glass has a low tensile strength but at great depths a glass sphere would be under very great compressive forces, and in these circumstances its strength would be more than adequate. In one experiment conducted by the Benthos Company in North Falmouth, Massachusetts, a glass float the size of a basketball was compressed in a water pressure tank to simulate a depth of 18,000 feet. A hardened steel pin was driven into the glass with a force of 3,000 pounds. The pin shattered and the glass remained undented. The military implications were demonstrated when forces of over 20,000 pounds from underwater explosions failed to break spheres of other ceramic materials. The major problem is in making the material predictable and making the surfaces tough for everyday life on the surface.

Recently acrylic plastics like Perspex or Plexiglas have become a favourite subject for tests. Acrylic is immune to the corrosive action of sea water and so are its glued joints. Therefore there is no need to anticipate and 'design round' stress corrosion or fatigue corrosion. These are conditions in which corrosion makes the tiny cracks resulting from fatigue much worse. Acrylic is therefore already being used for windows, and even large spheres have been made and tested. One American project NEMO involves the use of the plastic sphere as an aquatic free flight balloon which will go up or down in continental shelf depths by adjusting buoyancy. Since it is transparent in every direction, the view is far superior to that from a steel sphere with windows which costs six times as much.

MODERN BUOYANCY MATERIAL
density 40 – 50 lbs per cu. ft.

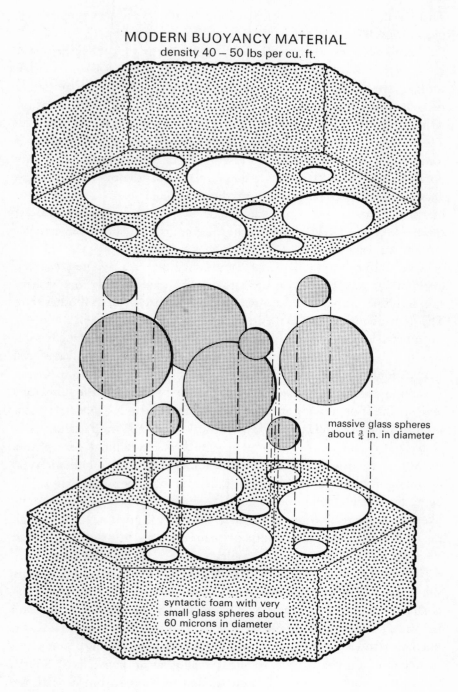

massive glass spheres
about ¾ in. in diameter

syntactic foam with very
small glass spheres about
60 microns in diameter

There is already one plastic submersible, the *Kumakahi*, operating to 200 feet in Hawaii.

Another bit of strange vocabulary that features in the submarine world is 'syntactic foam'. This is a designer's rescue material, for if a submarine turns out to be insufficiently buoyant, pushing blocks of the foam inside the outer hull but outside the pressure spheres will go a long way to restoring the situation. It consists of tiny hollow glass spheres set in plastic. The spheres are usually less than a tenth of a millimetre in diameter but sometimes glass spheres as large as an inch in diameter may be incorporated. The material weighs around 40 pounds a cubic foot, whereas water weighs over 60 pounds—it therefore contributes 20 pounds of buoyancy. It needs to be quite strong itself to withstand external pressures; it is made on both sides of the Atlantic in ever growing amounts wherever 'instant' buoyancy is needed.

Curiously enough, one of the most dangerous areas of present ocean exploration techniques is on the surface. The only notable loss of a research submarine happened as it was being launched from its mother ship. The deep-diving *Alvin* had no crew on board but a wave went down the open hatch and she promptly sank in 5,000 feet of water off Cape Cod. It was the end of the season and bad weather prevented her salvage until many months later. The task was accomplished with the help of Ocean System line and the *Aluminaut*. Getting the crew out of a little submarine and towing it or hoisting it on board sounds easy enough, but in fact it is both technically difficult and of critical economic importance. The Vickers *Pisces* for example cannot be launched in anything worse than a Force four wind which in the North Sea means substantial limitations on operating flexibility. Seas around American oceanographic centres are better but the difficulties encountered are still great. There is nothing quite like it in ordinary maritime experience. Ordinary submarines have enough range to go into their home ports by themselves, so the problem does not arise. Recovering or launching a lifeboat at sea involves comparatively slight weights, but submarines are usually over 20 tons. Lockheed's mother ship has a dock at the stern into which *Deep Quest* manoeuvres herself and is then lifted out of the water by a platform which comes up underneath the vessel. Other ships use A-shaped cranes which luff out over the water, but there is still the problem of 20 tons of submarine swinging about as it is manoeuvred on board on a rolling pitching small ship. The D.S.R.V. avoids the problem by being launched and recovered from beneath the waves but such solutions are impossible in the commercial world. This is another reason why the heavier deep diving research submarines have found making a living very difficult.

Commercially, the most successful submarines have been the less ambitious and the least expensive ones. Sun Shipbuilding, for example, have built the five ton Guppy type which works with an umbilicus in shallow water. It costs just less than 100,000 dollars. Another firm, Vocaline Air Sea Technology (V.A.S.T.) in Maine, acquired a whole company developing one-man submarines with its inventory for less than 100,000 dollars. Brown and Root has bought its own submarine from Perry Shipbuilders for pipe-line inspection. In most cases, no company wants to pay money for capabilities that are not required. If most pipe-lines are laid at depths less than 600 feet, no-one wants to pay extra for a submarine that will dive to 1,000 feet.

It is quite plain that there are two distinct classes of submersible apart from the military ones. There are those tailored to do a certain job in a predictable environment and the exploration craft. An extreme example of the latter, is the *Ben Franklin*, which was designed for exploring ocean currents by drifting along in them for weeks at a time. On its maiden voyage, it went up the east coast of the United States in the Gulf Stream meandering along at about 600 feet so that its six-man crew could spend four continuous weeks observing the current.

One thing that the more sophisticated exploration craft has done is to show the need for better sensors, viewing instruments, sonar devices, communications equipment, and remote manipulators or 'telechiric' devices. There is no doubt that the plethora of American submarines provided a market for such devices and big advances have been made; even so, bigger advances need to be made in the future.

It is commonly said that in the sea sound waves take the place of light, and over any sort of medium or long range this is broadly true. At short distances, however, light is useful and various electronic tricks can be employed to improve things. Even pure distilled water absorbs light and below 400 feet at sea, natural light has virtually disappeared. The last wavelengths to be absorbed are in the blue-green regions and most illumination systems try to use these penetrating wavelengths. If a light is shone ahead from a submarine, the effect is very like that of a car putting on its headlamps in fog, so much light is scattered back that except at very close ranges, the situation is worse than before. Putting a light next to the subject to be viewed is one expedient, but even then there is a diffuse scatter which reduces contrast badly. Here then is a chance for the device that used to be called the solution looking for a problem, the laser. There are several systems, some of which, like synchronously scanned laser T.V., are very difficult for even the most determined populariser to describe, but an idea of the sort of trick possible may be gained from the simplest method of all, a pulsed laser system with a gated receiver. It is simplicity itself, a pulse of blue-green light from say an argon laser is sent out. A very short pulse may be only a few feet long. As it goes out through the water, the receiver is off, so scattered light is ignored. When the pulse strikes the object, say a pipe-line valve, the receiver is switched on. Therefore, only the light reflected from the object is received. When the pulse of illumination has gone past the object, the receiver is switched off. So far as the receiver is concerned, only a thin slab of water is illuminated and if the pulses are fast enough, the images are combined and appear steady on a T.V. screen. The word 'gated' describes quite aptly the receiver's behaviour; it only opens the gate for the fraction of time the pulse of light is passing through the desired zone. The range of the zone can be selected as required. Effectively this system is as good as would be achieved by putting a light alongside the object to be viewed. It can be readily appreciated that devices like these are the first steps in a new technology.

The use of sonar (or Asdic) devices on the other hand has long been established in the military field. Indeed, without sonar sets, military submarines would be quite blind. A quick look at a bit of the detail of sonar shows that it too represents a distinct technology with a close resemblance to radar. There are the familiar irreconcilable problems. To get a sharp picture of a distant object, a very tight compact pulse of sound must be emitted. The higher its pitch or frequency, the greater the attenuation of the signal. Therefore, higher powers are striven for. Some of oceanography's high flying companies like E.G. & G. in America founded their success on reliable sonar devices that would produce a map of the sea bed alongside the ship's track much as a sideways-looking movie-camera does from a low-flying aircraft. One of the best pieces of ocean hardware produced in Britain is the large towed sonar device called 'Gloria' which was developed over many years by the National Institute of Oceanography in deliberate isolation from the Admiralty Research facilities so that problems of secrecy would not arise.

Another important area of development is that of telechiric (remote hand) devices or manipulators. Many devices have been modelled on those used in atomic energy to handle radioactive materials; the simpler ones are comparatively weak and lack 'touch'. Another problem is that power is transmitted to the 'hands' by hydraulic hoses which deform to the outside pressure. The General Electric company in the United States which has sat outside the ring until recently, has an important corner in this area. Like General Dynamics, it has already developed a manipulator without hoses so that external pressures are of less account and vulnerability is lessened. It also has spin-off from a military project in which an 'iron man' was designed to augment the muscle power of an operator standing inside it. The device uses feedback so that the operator is able to sense how strongly his remote hands are holding onto things. If any complex assembly has to be done on, for example, a mid-ocean ridge military base, small submarines with such sophisticated manipulators would be invaluable. With the simple manipulators adapted from atomic energy models, the time taken to do a job has been literally a hundred times longer than that taken to do the same job on a bench. *Aluminaut* took 18 hours to push an expanding bar down the sunken *Alvin*'s open hatch and then hook the bar on a cable during the salvage operation.

Not only are submarines clumsy and shortsighted, but they are also very bad at knowing where they are. A common technique is direction from the surface assuming the mother ship knows where she is. She keeps her charge in view on a sub-hunting sonar set, plots his position and tells him over an acoustic telephone. This latter device has a long military pedigree; a sound sonar emission is modulated by a human voice just as a carrier wave is in radio. Sophisticated submarines like *Deep Quest* can give themselves three point fixes by monitoring sound-emitting beacons placed on the bottom in positions determined by the mother ship. Another useful device is doppler sonar in which the speed over the sea bed is computed by the rise or fall in pitch of a sonar emission bounced back off the bottom. The principle is the same as in the police speed-trap radar, speed can be read off directly and when combined with a gyro or magnetic compass heading a form of dead reckoning can be achieved. Most of these systems are greatly in need of improvement.

Last of all comes the problem of power. Virtually all research submarines except the U.S. *NRI* which is nuclear powered, have to rely on batteries. The important figure is the weight of power source per kilowatt-hour. It is here that the advantage of having a power source on the surface feeding through an umbilicus is most marked, and for hard work the self-contained submarine is much inferior to devices like the Cammell Laird Sea Bed Vehicle. The bigger the battery, the bigger the vehicle needed to carry it, therefore, the greater the demand for power and so on in a vicious spiral. Which battery is used depends on cost. If it is worth paying for it, silver zinc batteries will give 35 watt-hours per pound weight; familiar lead acid car batteries (by far the commonest power source) will give 15 watt-hours. The problem is similar to that met in space and the fuel cell may again be the answer for missions lasting no more than a few days. In Sweden, work has begun on ammonia oxygen systems for conventional submarines. The ammonia is broken down to produce hydrogen which is combined with oxygen to give current directly. Other work at Electric Boat has used hydrazine and oxygen and at General Electric hydrogen and oxygen have been generated from sodium aluminium hydride and sodium chlorate. This device was said to produce 175 watt-hours per pound. Similar figures have been achieved by Energy Conversion Ltd. in Britain using the zinc air battery. The major snag with most of the systems using fuel cells is that they need to be inside the pressure sphere. The lead acid batteries on the other hand can be surrounded in oil and allowed to function at the ambient sea pressures between the man-carrying capsules and the outer hull.

The similarity between oceanography and space exploration extends as far as the man or machine controversy. Just as there are two schools of opinion for and against man in space so there are two schools of opinion for and against man in the sea. Many people believe that remote controlled instruments can do anything that 'inner space ships' can do and do it more cheaply. People holding these views tend to be scientists and as in space they are opposed by the military and the technologists. Some of the remote-controlled instrumentation, however, is getting every bit as technologically ambitious as the expensive submersibles. A good example is the work of Frank E. Snodgrass at the Scripps Institution in California. He has constructed capsules to operate on the ocean bottom which are accurate enough to measure the ocean tides directly. In the ocean, tidal waves are measured in inches. It is only on the shallow continental shelf that the big, slow, twice daily waves show up as rises of 10 feet or more. However, measuring a rise and fall of inches from several thousand feet down on the sea bed, and abstracting the figure from the millions of short period surface waves is a tall order for any instrument.

THE DEEP SEA INSTRUMENT CAPSULE

This instrument can be called to the surface when it is 5000 m. down on the ocean bed. It can remain on the ocean bed for several months.

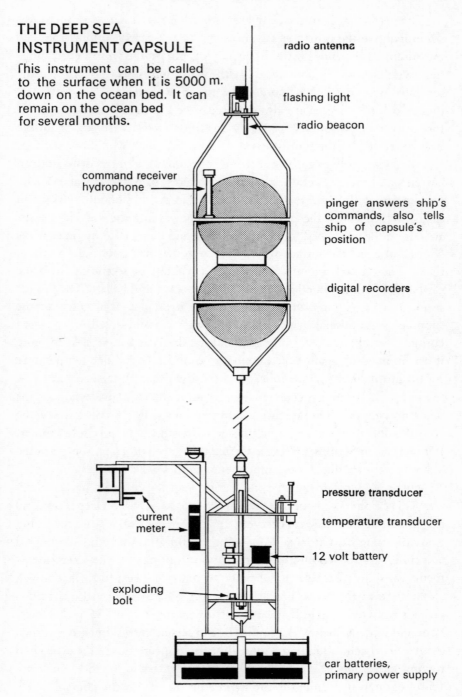

radio antenna

flashing light

radio beacon

command receiver hydrophone

pinger answers ship's commands, also tells ship of capsule's position

digital recorders

pressure transducer

temperature transducer

current meter

12 volt battery

exploding bolt

car batteries, primary power supply

The performance of the capsule is astonishing. The height of the water column above the instrument is measured to the nearest millimetre. The measurement depends on the frequency of vibration of a stretched tungsten wire whose tension varies with water pressure. To correct the readings temperature measurements to an accuracy of at least one thousandth of a degree are also required. The Hewlett Packard temperature instrument is so sensitive that it has to be specially isolated from the heat generated by the transistorised electronics, which would normally be regarded as negligible. The Scripps instrument also records bottom currents and their direction, a particular interest of Mark Wimbush who collaborates with Snodgrass. He is interested in the thin boundary layer on the ocean bed where the earth's heat is conducted into the sea. The sensors are left on the ocean bed for six weeks or so and then called up on acoustic command from the surface by a research vessel that goes out to collect them. They detach themselves by explosive bolts or solenoids from the car batteries which function as both power source and ballast. The project seems to have begun as the usual laboratory rig which, if it didn't use the supposedly traditional string and sealing wax, certainly used off-the-shelf components like the recording gear and standard car batteries. In the end, it has bumped up against the boundaries of technology. The temperature measuring device, for example, is so sensitive that allowance has to be made for the heating generated by the momentary compression of the water flowing past it at say, three centimetres a second. Needless to say, the recorded information has to be interpreted with the help of a computer. It is said that at Scripps, pulses beat faster as the deep ocean sensors go in the water because so much of the annual budget goes with them.

There is another unbelievably complex device built under Navy auspices at Point Loma near by in San Diego. It is called 'Deep Tow' and is an instrumented steel fish carrying magnetometers, sonars and recording gear. By using capacitors and operating for only thousandths of a second at a time, it is able to push out large amounts of power, the sonar instruments can thus achieve a useful range to map the ocean bed. It is towed along close to the ocean bed at 20,000 feet where the pressure is 10,000 pounds per square inch. It measures its own true height above the ocean floor and its depth below the surface once a second by an upward and then a downward echo sounding. Control instructions go down the armoured towing cable and it is said to be so complex that only three men are skilful enough to handle the thousand adjustments on the control panel.

THE DEEP TOW VEHICLE
Used by the Scripps Institution it maps the
small scale topography of the ocean floor.

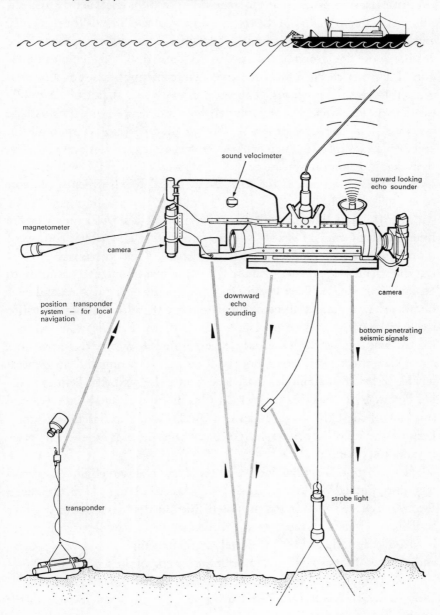

sound velocimeter

upward looking
echo sounder

magnetometer

camera

camera

position transponder
system — for local
navigation

downward
echo
sounding

bottom penetrating
seismic signals

transponder

strobe light

Over the last seven years, three have been built for less than a million dollars each, or so they claim, and so far they haven't lost one.

A visit to either of the workshops where these instruments are made confirms the impression that the standard of technological achievement is fully comparable to that of the space industry. The demands for reliability and robustness in particular are very pressing. If spending in space is to be defended on the grounds that doing technologically difficult things is good for America, then the same argument can apply with possibly even greater force in the oceans. The only drawback is that from a public relations point of view, both submersibles and remote instruments lack glamour. Unlike a space capsule, they do not produce sharp beautiful colour photographs at which ordinary voters can express amazement. The best that oceanography can do are the 'houses under the sea' projects like Cousteau's Conshelf, the U.S. Navy's Sealab experiments, or most recently, Tektite II.

Tektite I and II were both set up by the civilian Department of the Interior with General Electric as prime contractor. The habitat consists of twin large chambers standing on a ballast base. The depth is only 50 feet but the scientist aquanauts who are the experimental subjects spend up to 60 days in residence. They breathe oxygen and nitrogen mixtures and their bodies are saturated at ambient pressure, they therefore cannot return to the surface without decompression. Much of the data is about psychological matters; who gets on who's nerves and how quickly; how does social class affect social stress; does wall-to-wall carpeting and the general feeling of being in a Hilton Hotel at someone else's expense help to make life better. It all looks rather nice on film lit from above by the tropical sunshine of the U.S. Virgin Islands and, as an added attraction, six members of the crews are women. As one magazine put it, the 'undersea frontier tests man's resilience'.

Some people have an idea that the U.S. frontier could be advantageously pushed out to sea, to the Mid-Atlantic Ridge in fact. General Electric, Tektite's maker, has proposed Bottom-Fix, described in a glossy brochure in which all the men are in the inevitable shirt-sleeves. The idea is to build a modular base of several interconnecting 'Pyroceram' spheres and a 12,000-foot capability is forecast for the end of the seventies.

That completes an unavoidably condensed survey of the more notable projects which have contributed to knowledge of what goes on in the oceans. It is clear that in spite of a great deal of effort, man's capability to do anything beyond the continental shelf is severely limited. There are only two or three manoeuvrable submarines that dare venture even half-way to the ocean floor, and in spite of progress with underwater cameras, magnetometers, sonars and the like, anything that falls to the bottom of the ocean still stands a very good chance of staying there. Only if it is of vital strategic importance, like a lost H-bomb or a nuclear submarine, is it worth while deploying expensive but still clumsy devices to rummage in Davy Jones's locker.

14

Marine Mines

IN AMERICA THERE ARE SEVERAL PEOPLE WHO SEE THEMSELVES COMPETING FOR a special place in the history books of the twenty-first century, if as seems inevitable popular history writing stays the same. Just as today we read that Bessemer began the modern steel industry and Sir Frank Whittle was the father of the jet engine so in 2050 they hope that they will be called 'the father of ocean mining'. It would be invidious to name those in the running for this honour, indeed most people outside the oceanography circle would regard the title as a mark of eccentricity. There are no ocean mines actually being worked, indeed there hardly exists the equipment to map out such mines if they really are there.

Nevertheless the projections and plans are detailed and extensive, all that is needed is money. In other parts of the world the American enthusiasm for minerals on the sea bed is regarded with something approaching amusement or condescension. In only one other country is the prospect taken seriously and that is because that country, Japan, is unique in being a major industrial power virtually without natural resources. The Americans for their part have more than enough minerals to feed their own economy, with the major exceptions of manganese and tin. Manganese in particular is a strategic material because of its essential use in steel. At one time something over a half of America's manganese came from Russia and these supplies were cut off in the early fifties when the cold war began. Another supplier of manganese in the past was Cuba. Americans may therefore be forgiven for feeling a little sensitive about manganese and it was an American mining engineer, John L. Mero, who pointed out in 1952 that his country could, if it chose, get all the manganese it wanted from the ocean floor. The source would be the well-publicised manganese nodules.

These nodules were discovered a century ago by the *Challenger* Expedition (1873-76) and almost every oceanographic expedition since has brought them up from somewhere or other. They are found all over the Pacific and Atlantic Oceans, they have even been found close inshore in Scottish lochs. Continental shelves and deep trenches alone seem to have only small quantities. In the Pacific Ocean as a whole Mero reckons that 1·6 million million metric tons would be a reasonable estimate. The average percentage by weight of manganese dioxide is 31 (calculated on the *Challenger*'s specimens) and on land this would count as a fairly rich ore. In some regions the percentage can be as high as 50. Some nodules are high in even more valuable metals like nickel and copper.

The amount of excitement that manganese nodules inspire in everyone except practical engineers, bankers and mining houses has recently been exceeded by the brine sinks. These are deep pools of hot metalliferous brine in the Red Sea which are floored by thick deposits of metal-rich sediments. The discovery of brine sinks has proved so intoxicating that German companies have actually hazarded substantial sums of money in attempts to see if they can be exploited. The best nodules and the best brine sinks are both deeper than 6,000 feet, and somehow nature has arranged things so that the easier the minerals are to get at, the less is the effort worth while in economic terms. The zinc, copper and nickel are deep, whilst the minerals closest to the surface are sand, gravel, phosphate; alluvial tin of course is an exception which is derived from deposits on land and happens to lie underwater close inshore.

Perhaps the best way to tackle the many-sided subject of ocean minerals is to look at the nodules first, then the brine sinks and then switch abruptly to two practical exercises in ocean mining neither of which have made their sponsors rich. In this way reality will temper the infectious optimism which it is particularly easy to contact in this branch of oceanography.

The group with the most substantial capability in the ocean mining area is certainly American and is probably Deepsea Ventures Inc. Unlike other advanced ocean technology companies like, say Ocean Science and Engineering, Deepsea Ventures seems to be interested exclusively in minerals. It is a subsidiary of a huge four-billion-dollar Texas chemical company called Tenneco. Its remit from the parent company included other areas of ocean exploitation, but Deepsea Ventures' President, John E. Flipse, decided on a deep ocean mining programme because he thought it would pay out soonest. Flipse is a large muscular man with a crushing handclasp who looks as if he has only recently given up some physically exhausting open-air pursuit like oil drilling, or maybe boxing. Some of the mental qualities appropriate to both those pursuits are likely to be required of him in the next two or three years.

Deepsea Ventures has a low block of offices inland from Newport News south of Washington D.C. It began as an attempt by the Newport News Shipbuilding and Drydock Company to diversify in 1962. That shipbuilding company, which might be called the John Brown's of the United States, has always had a very high reputation for quality and had a great deal of experience in engineering for the U.S. Navy. In 1968 both Newport News and its onetime diversification were under the wing of Tenneco. Although the giant parent could finance a major project itself, and cynics would say that if there's anything in ocean minerals it ought to, the chosen mode of operation is to get together a consortium to share the risks.

The problem is, as they like to say, multifaceted. For a start manganese ore refinement techniques suitable for use on land ores are not likely to be appropriate for the weird nodules. Flipse makes encouraging remarks on this score and says that sufficient work has been done to make sure that effective refinement techniques will be ready when the nodules are. Another problem is that to be economic an ocean mine would be so big as seriously to disturb the world's metal markets. It would, for example, provide 40 per cent of the total U.S. consumption of manganese, and 50 per cent of its cobalt. Since the U.S. is the largest consumer of these metals, the impact of an ocean mine on the economics of the developing countries which presently produce the metals may well be serious. From an economic point of view, however, it is nickel and copper which are likely to be most profitable to ocean miners and the quantities of these metals would not be so embarrassingly large. An ocean mine would be simply an area of the sea bed about 30 miles square covered with nodules of suitable size and composition. Sometimes the nodule deposits become joined together and are literally pavements. Although pavements of manganese sound attractive, with an echo of those legendary places where streets were paved with gold, an engineer would regard pavements as unsuitable simply because they cannot be dredged. Ideally the nodules need to be the size of small potatoes, two inches or so in diameter.

It may be guessed that in terms of equipment at any rate the problems are even bigger than those that face the oil drillers. John Mero in his book *The Mineral Resources of the Sea* sets out the possibilities. The critical difficulty is depth. At any depth beyond 2,000 feet a conventional clam-shell-type bucket would spend so long going down and coming up that it would need to be of enormous capacity to be economic. The problem is exacerbated by the fact that the nodules are probably only one layer thick so the bucket jaws would have to open extremely wide.

Other conventional types of dredger are out of the question. The popular cutter suction type for example mobilises bottom sediments by pushing a rotating drill into the bed and sucking. Its maximum operating depth is rarely over 150 feet. The old-fashioned ladder-of-buckets design is similarly limited. Indeed, there seem to be only two or three possible types. The simplest is a deep sea drag dredge, an enlarged version of that used by oceanographers. Deepsea Ventures has collected most of its nodules that way, in one-ton lots. Mero envisages a 2,000-ton ocean tug type of ship with a vast barge alongside to receive up to 5,000 tons of nodules. A dredge bucket 20 feet long with a 12-foot mouth carrying 13 tons per haul would be used. It would be fitted up with underwater T.V. and a sonic pinger to tell the operator when it had reached the sea floor. Mero gave detailed costings in 1964; mining costs came to around 20 dollars a ton of nodules in depths of 5,000 feet. This figure assumed a density of nodules at two pounds per square foot which is common enough. Total capital costs were estimated at two million dollars. In 1969 Mero said he still stood by those figures with due allowance for inflation and other rising costs. One fundamental limitation is the fact that the maximum speed at which the bucket can be hauled up to the surface after filling is likely to be 1,000 feet per minute, because the power required to overcome the water resistance rises as the cube of the velocity. According to Mero's calculations, although the capital costs are less and the technological unknowns relatively few, this system is substantially inferior to the system that has become known as the underwater vacuum cleaner.

The main reason is that the vacuum cleaner system gives a continuous operation.

The vacuum cleaner system is basically a hydraulic suction dredge, a pipe hung over the side of the ship reaching to the bottom. As the ship steams slowly along a pump sucks up sand and water. Unfortunately, such a simple device will not work in depths greater than 33 feet because of the fundamental laws of hydrostatics. The maximum suction that can be developed is only one atmosphere if the pump is on board a ship on the surface. One atmosphere is equivalent to 33 feet or so of sea water. The answer is to submerge the pump, and immediately the problems begin. Mero's original suggestion was to suspend the pump from a stabilised float tank submerged below the influence of the waves. The control ship would be connected to this by flexible pipes and cables.

However, Deepsea Ventures' prototype dispenses with the stabilised submerged float and opts for using a conventional ship and operating through a 30-foot square well, cut in the mid-section and down through the keel. In this well is an elevator which can travel 40 feet or so above and below the main deck. On it the dredge, with its 6,000 feet or more of pipe, and the heavy pump is to be assembled section by section with the ship rolling no more than five per cent. It can readily be seen that the rigging problems will be just as difficult as in the oil business and the only consolation in sight is that round trips ought to be less frequent. Flipse has in mind an 18,000-ton deadweight ship capable of holding up to 15,000 tons of ore for transhipping into ore carriers. These would ply continuously taking the nodules to a shore refinement or smelting plant every four days. Ore would be transferred along flexible pipes whilst dredging continued. Nodules would be pumped from the bottom at rates of at least 150 tons an hour in a continuous stream with a fluids to solids (i.e. seawater to nodules and sand) ratio of 25 to 1. The nodules would rattle up the 20-inch pipe at a fair pace, 15 feet a second and, needless to say, great pains have been taken to ensure that failures will not result in disastrous blockages. Various automatic dumping mechanisms have been designed and tested. The dredge head itself is a sort of massive rake pivoted on the sea bed and connected through a long rigid handle or truss to the end of the hollow steel piping. Here again is another pivot or hinge, the pipe being held down by a dead weight. The pump capsule will be submerged to 500 feet or so and pressurised with air to stop water leaking in. It is all rather unusual engineering and much of it has to be purpose built; apparently it is not

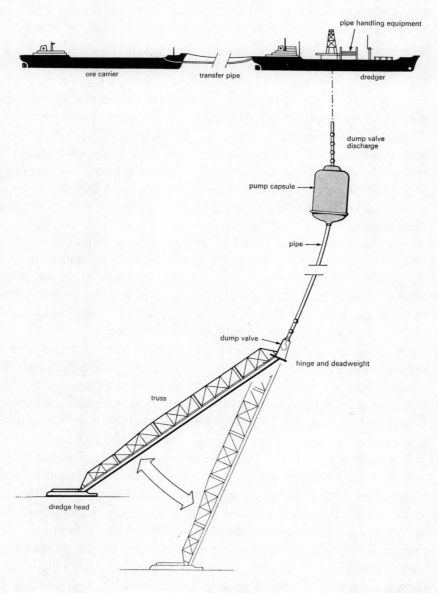

A DEEP OCEAN MINING SYSTEM
A design approved by
Deepsea Ventures Inc.

possible to borrow even the pipe designs from the oil business, new joints
have had to be designed and patented. The automatic dump valves, the
hinges, the pumping unit, the pipe stowage equipment and several other
sections have all been patented.

The various designs of rake have been evaluated in model form and
tested in tanks. Essentially they have to sort out nodules of the right size
from those too big or too small. At least half the water used to transport
the nodules up the pipe comes from clear areas above the dredge head.

Although some of this technology is in the form of paper studies and
computer tests, it is important to realise just how much solid work has
been done. When Tenneco first bought up the company from the Newport News shipyard (the shipyard purchase came later) they are reported
to have paid several million dollars for the various assets, the patents and a
small research ship the R.V. *Prospector*.

One gimmick Deepsea Ventures has successfully employed is a free
manganese nodule assay service. They will analyse any nodule free of
charge for anybody in exchange for the precise latitude and longitude
whence it came. In this way Deepsea Ventures has amassed what it claims
to be an unrivalled knowledge of the distribution of manganese nodules,
and the company claims it has a bank of 30,000 data points.

A certain air of discrimination about the best locations for nodules is
somehow one of the most convincing aspects of the whole affair. No-one
talks in terms of broad principles, one gets the feeling that there are
certain areas marked on securely locked away charts where all the various
factors are just right. The bottom is clear, free of obstruction, the nodules
are abundant enough, of the right size and the right composition, the
surface weather and sea states are good enough and the haul to a shoreside
refinery is short enough. Unfortunately the nodules richest in nickel and
copper seem farthest from land, in equatorial mid-Pacific. The nearshore
areas off Japan and California are manganese rich but the shallower the
water the less valuable the nodules become. On the Blake Plateau area off
Florida, the nodules, although often less than 3,000 feet deep tend to be
high in iron and calcium carbonate and very low in valuable metals like
copper and nickel. Nevertheless, Deepsea Ventures had chosen this area to
demonstrate its technology at its own expense in 1970.

The company purchased a 320-foot cargo vessel of 7,500 tons and
renamed it bravely Research Vessel *Deepsea Miner*. After looking at
market trends in the basic metal industry, Flipse named 1974 as the target

year for deep ocean mining and metal production. A little later it was announced that Tenneco and Metallgesellschaft A.G. of West Germany were linking up for the evaluation phase and *Deepsea Miner* set off to prove the prototype system on the Blake Plateau.

John Mero, who worked with the East Coast team for some time, left to run his own consultancy, Ocean Resources Inc., on the west coast of America near San Diego. His offices are in an estate set in a dry valley and every factory around has a sign saying something that ends in 'itonics' and sounds highly advanced. Westinghouse's Ocean Laboratory is down the road, Scripps is over the hill and the city of San Diego, bidding to be the oceanographic centre of the world, is only ten miles away. Mero was educated as a mining engineer well inland in North Dakota but he got hooked on the sea as an undergraduate. He radiates a controlled but determined optimism about ocean mining. His confidence is infectious and founded on long experience and deep study. 'If copper costs a dollar a pound then copper from the sea will only cost 20 cents. Nickel will be so cheap that it will make stainless steel possible even for structural purposes. There will even be stainless bridges – no more painting.'

Mero thinks on a time-scale of 20 years. One factor which might precipitate the big mining companies to invest sooner is the likelihood of the developing nations demanding more money for their minerals. But ocean mining he asserts, is commercial right now and has all sorts of advantages, there will be no pollution, no disfigurement of the landscape, no labour problems, no politics, everything can be capital intensive, automated and sure. And there is enough of the rarer but essential metals at the bottom of the ocean to keep the world going for thousands of years. The enthusiasm is infectious whilst you are in the United States, but somehow, like so many other enthusiasms, it tends to wear off when you get back to Europe.

The most extraordinary thing about the manganese nodules is that nobody even now has a clear idea of how they were formed, or why they are where they are, on the surface of the sea bed. Sometimes they are found to have grown round shark's teeth, fish bone or bits of pumice stone, coral fragments or naval shell shrapnel. The most common nucleus is volcanic material. They can be round, knobbly or tablet shaped, big or very small. The largest, brought up with a telephone cable from the Pacific, weighed nearly a ton and not surprisingly was dropped back into the sea. The largest one recovered and kept came from the Blake Plateau and weighs over a hundred pounds. Frequently where whole pavements of nodule material form the outcropping sea bed rocks are covered with it. Individual nodules have a sort of onion skin structure and are clearly concretionary in nature; somehow they are built up layer by layer from the surrounding sea water.

It is very difficult to deduce the mechanism of formation partly because it is so slow. Radioactive dating has shown that it can be as slow as 0·01 millimetre in 1,000 years. Nodules are nearly always found on the surface of the sea bed and this is itself a mystery because they can form in sediments which are being deposited faster than the nodules are growing. Therefore they ought to be buried. One mechanism put forward is that they might grow on top and dissolve underneath; that is, form only in the interface between the sea and sea bed. Another reason they are found on the surface could be that they are simply pushed there by worms. Why do they have so much manganese? Why in some regions do they concentrate nickel and copper? Goldberg has suggested various mechanisms which depend on the electrical properties of colloidal particles. Sea water is normally saturated with manganese and iron so as more comes in from rivers, some ought to precipitate out in very fine particulate form, that is in a colloidal form. Exactly where this comes to rest depends on the electrical properties of absorbent surfaces, and good conductors like fragments from naval shells are known to accrete manganese and iron oxides much faster than non-conductors. One naval shell 50 years old had accreted three centimetres of nodule material. Many scientists are nonetheless unconvinced by the colloidal electrochemical theories and invoke bacteria instead. One thing can be stated with certainty, that manganese nodules will remain the source of dozens of oceanographic Ph.D.s in years to come.

From an economic point of view there can be little doubt that the nodules will be mined sooner or later and that they will be a valuable source of manganese, cobalt, nickel and copper. The most likely countries to begin the effort are Japan and the United States and it may well be the former that makes the going by starting a Pacific Ocean project within the next five years.

So much then for the nodules with their faint air of unreality. For some reason the very word nodule makes people smile, perhaps because in English culture it has an echo of Enid Blyton's Noddy in Wonderland; it is also an affectionate term of abuse.

The other extraordinary features of the ocean bed have an even more suspect name, the brine sinks. If unbelievers can poke fun at dredging nodules they can be even more scornful of pouring money into brine sinks. Yet these features may prove to be of major economic and geological importance, they may even be of fundamental importance in understanding the processes whereby mineralisation occurs on the continents themselves, truly the primary ore factories of the planet.

Brine sinks were first discovered in the Red Sea west of Mecca. A number of oceanographical ships had commented on the high salinity and the high temperature of water near the bed of the Red Sea. It was assumed that salt water near the coastal shallows had been warmed up and had suffered evaporation in the strong sunshine. This denser brine was assumed to drain inwards to the bed of the sea. In 1964 the British ship R.R.S. *Discovery* was mapping the sea bed when the sonar apparatus registered a major target at 5,000 feet down. Something which was not the sea bottom gave back a strong echo. It evidently corresponded with some sharp change in density in the sea itself. A temperature gauge, a bathythermograph, was run down and the temperature below the interface was discovered to be an astonishing 111°F., at least 40°F. hotter than the rest of the sea. Hot water normally floats on cold but this, whatever it was, did not. The only explanation was to assume that the liquid was a brine so saturated with dissolved salts that even when 40° hotter it was still denser than sea water. Samples were collected and it was found that the brine had a salinity of 286 parts per 1,000. The normal salinity even of the saltier surface waters is only 38 parts per 1,000. Clearly the notion of warming and evaporation in coastal shallows could not account for this. The hot brine appeared to lie in a pool or sink eight miles square, thereupon dubbed the '*Discovery* Deep'. Free fall corers were deployed to sample the bottom. These are supposed to plunge down, take a core and release a part of themselves carrying the core and this is then supposed to float back to the surface. None of them came back up however. The *Discovery* sailed away and, as is customary among scientists, told the rest of the world about it.

Next year the Americans from Woods Hole came in their ship *Atlantis II* but whilst preparing to investigate the *Discovery* Deep, their ship drifted five miles to the north. Astonishingly, they found another separate brine sink, thereupon called '*Atlantis II* Deep'. This proved to be even hotter, 133°F., and bigger, the sediment which was finally sampled proved to be too hot to touch, it looked like black tar. Some of the metal concentrations found in it were remarkable, zinc was 1,500 times more plentiful than in sea water and lead was 30,000 times more plentiful. Something other than evaporation lay behind these features. Evaporation increases the content of highly soluble compounds like common salt or potassium chloride but the Red Sea brines, to say nothing of the sediments beneath them bore no relation to brines like those of the Dead Sea which are formed by evaporation. Nor did the pattern of enrichment have any connection with the manganese nodules.

The brine sinks might seem at first sight to be a local freak. Indeed one early theory was that ground water from Saudi Arabia had percolated through coastal lead and zinc deposits and evaporites so becoming brine, had run down through local volcanic areas and become hot and then come up in the sea bed. The really curious thing about the sinks, however, is that they lie dead in the centre of the Red Sea on the prolongation of the line of the mid-ocean ridge which runs through the Indian Ocean. As we have seen it was becoming widely accepted in 1964 and '65 that mid-ocean ridges have a rift valley in their crests, that they are the site of basaltic dyke swarm emplacement and are generally speaking an area where the earth's crust is splitting or being split apart. It became clear that the Red Sea is the actual tear line between two separating continents. The rift valley in the centre of the Red Sea, if such it is, links with the rift valleys of East Africa and the Dead Sea farther north and is associated with earthquakes and vulcanicity. There has been a suggestion that Moses' celebrated flight from Egypt in which a great flood came up the Red Sea to engulf the pursuers of the Israelites, took place at the time of a vigorous earthquake. The Book of Exodus refers to a 'pillar of fire and of the cloud' and the retreat of the sea. Could this be a distant volcanic eruption and associated movements of the sea floor?

Anyway crustal activity and vulcanicity is a fact and the brine sinks now take on a great significance. They evidently lie in the rift of a mid-ocean ridge where it runs into a continental land mass. Farther back down the ridge and farther back in time, therefore, there might have been other brine sinks floored with rafts of metal-rich sediment. What happens where other arms of the mid-ocean ridge run under a continent today? One such place is Southern California – what happens there?

In the area is a large inland lake called the Salton Sea, and nearby are several hot brine springs. The whole area lies in a rift valley system with

high heat flow. American engineers have drilled several holes to tap geo-
thermal steam supplies for power generation and industrial use. To prove
the reserves the usual procedure is to allow the steam to blow off to the
atmosphere for several months; if there is then no sign of a loss in pressure
one may assume that the source is dependable. In one hole the pressure
stayed high but the flow rate was reduced to a trickle and it was assumed
that something was being deposited in the pipes. When the deposit was
extracted it was found to contain silver ore which had been deposited at
the rate of 1,500 dollars worth a day. Other metals in the deposit were
iron, copper and antimony. The steam, when separated from the brine, is
now used to generate about 10 megawatts of electricity, but it may well
be that the extraction of minerals will prove to be more profitable in the
long run.

With these remarkable examples of the power of brine to extract
metals from rock, geologists began to pay more attention to the whole
subject of hydrothermal systems and mineral deposition. Until oceanography had focused attention on the mid-ocean ridges and the newly discovered brine sinks, hydrothermal systems had been just another bit of
geology that was quite interesting but lacked any particular significance.
Now hydrothermal specialists find their work absolutely central to
economic questions like: are there other fossil brine sinks? and to
academic questions like: is our theory of mineralisation wrong or at least
badly distorted?

Deposits from the hot brines in the Red Sea and the Salton Sea are
undoubtedly of economic grade. The best drill pipe in the Salton Sea
produces three tons a month of a siliceous scale which contains from 10 to
40 per cent copper, one to six per cent silver and from half to one per cent
antimony. The Red Sea sediments contain up to 21 per cent zinc oxide
and four per cent copper. A recent American estimate reckons the upper
30 feet of the *Atlantis II* Deep sediments are worth two billion dollars
where they lie and the sediments are at least 300 feet thick.

One of the world experts on hydrothermal systems is A. J. Ellis, a New
Zealander whose home country provides many interesting examples,
though these are not apparently associated with a mid-ocean ridge. He has
performed many experiments exposing rocks to brine at a variety of
temperatures and pressures to see what goes into solution and when. The
solubility of minerals in the rocks depends on temperature and pressure
dependent chemical equilibria, and achieving the equilibrium conditions

i.e. when no more mineral will dissolve or precipitate out from the solutions) is difficult because it can take so long. It may need months for example to recrystallise some aluminosilicates from water colder than 400°C. Nevertheless, sufficient work has been done to reveal some unexpected results. Basalt and ordinary water at 350°C. and 500 atmospheres pressure produce a brine with 390 parts per million of chloride ion, 175 of sodium and 87 of potassium.

Such a brine is not particularly concentrated, it is not in the same class as the wholly saturated brines of the Salton Sea and Red Sea sinks which have salt concentrations of two or three orders of magnitude greater. Nevertheless, it is clear that brines can be formed and, once in existence, it is easy to suggest various ways of concentrating them until they are saturated. In the presence of really concentrated brines, Ellis has shown that the common rock type andesite for example can be easily leached to provide copper and lead. The result is a metal-rich brine which has nothing to do with so-called magmatic solutions or 'exhalations'.

The classic theory of mineralisation works very differently. It is based on the observed association of ore veins with granite and the classic area is Devon and Cornwall. The central idea is as follows; the liquid granite magma is intruded into the country rock and cools. As it does so, minerals crystallise out from the melt and the first ones to solidify are those whose composition is in equilibrium with the melt at high temperature. As the temperature falls, the remaining liquid is progressively impoverished in first calcic minerals, then sodic minerals, then potassium minerals; these minerals are usually felspars. Towards the end of crystallisation the remaining melt has lots of silica, steam and fugitive elements like fluorine, chlorine and metals like tin, silver and copper. The amount of remaining melt is very small in comparison to the vast quantities of magma which make the extensive tracts of granite in South-West England. Nevertheless, in the last dregs is all the tin, copper, etc., once present in several cubic miles of liquid rock. This final hot liquor is then imagined to force its way into any lines of weakness in the surrounding country rock and freeze to form the classic primary ore veins.

Geologists themselves were always fairly humble about their theories of mineralisation and the official geological account of South-West England retails the theory in a slightly diffident style that reflects the confidence with which the classical view was held. 'Vapours and solutions arising from the solidifying granite gave rise to the ore bodies. The emanations probably consisted in great part of stannic (tin) fluoride, carbonic acid, boric acid, and sulphuric and other acids. Stannic fluoride is supposed to have reacted upon steam to form tin oxide and fluoric acid, which in turn attacked carbonate of lime to form calcium fluoride or fluorspar.' Nevertheless, the association with granite is very strong and the minerals even lie in zones concentric with the granite. Next to the granite is tin, then comes tungsten, then sulphides of copper, then zinc, then lead, silver and antimony. The farther away from the granite the lower the presumed temperature of formation.

Since the discovery of the brine sinks, however, an alternative mechanism for mineralisation must be considered. Conceivably a saturated brine could concentrate tin, zinc and similar metals directly from solid basalt, and the whole lengthy path of concentration by fractionation of molten rock could be avoided. In fact, there are two alternative mechanisms. Either hot brines can leach country rock on land or they can leach hot fissured basalt in the mid-ocean ridge under the sea. In the Red Sea it seems that the metals do come from basalt. The latest American work on the Red Sea brines indicates that sea water has migrated for hundreds of miles along the fissures. The proof is too complex to detail here but in brief the concentrations of certain isotopes of oxygen and hydrogen were used to trace the origin of the water. Assuming that there have been no rapid changes in the isotope profiles of Red Sea water, it appears that the water came from the southern end near the Indian Ocean and has taken at most a few thousand years to make the journey of six or seven hundred miles.

An important question is that of how the sea water became a saturated brine. If it is required to go through an evaporite to become a brine, then active brine sinks must not be looked for in the deep ocean. This is because it is presumed that salt beds are formed by sub-aerial evaporation on land. It is also quite possible that in the Red Sea the brines became saturated because they dissolve salt from nearby sedimentary rocks. Out in the deep ocean miles away from land, this theory supposes that there would be no convenient source of solid salt to help the brines to be concentrated. Therefore brine sinks may not be common at all. Active ones will be close to land, fossil ones will have happened only where a continent was being split. One fly in this theoretical ointment is that ocean bed salt domes are turning out to be more common than ever suspected. Nobody knows much about them yet but there are signs that they can form in deep water at the early stages in ocean formation when the water circulation system is poor.

Alternatively, sea water could become concentrated into brine in the mid-ocean ridge fissuring, though exactly how is none too clear. Excess water would have to be driven off as steam and it seems likely that submarine vulvanicity does involve steam venting in the sea bed. Geologists have used words like exhalations and emanations for years in efforts to explain various effects so they should not object too strongly.

Clearly the least complex role of hot concentrated brines in mineralisation may be just to scavenge the base metals like tin, lead, zinc and copper from either intruded granites or the surrounding country rock. Mineralisation would then result by the brine intruding into fissures and it could well look like the pattern of mineralisation found in Cornwall.

However, the brine sinks may have a much more fundamental significance if they are found to be widely distributed along the 40,000 miles of mid-ocean ridge. Sea floor spreading may have split fossil brine sinks in half; obviously once the sink is physically destroyed the brine drains away but not the sediment layers beneath. These fossil brine sink floors may still exist, lightly buried in red clay, ooze, or even manganese nodules, some distance away from the ridges and symmetrically disposed on either side of them.

If the latest theories like the Dewey-Horsfield theory are correct and drifting continents are indeed welded together when oceans close and plates of crust have been forming at ridges and disappearing at trenches for at least 3,000 million years, then one must pose two questions. What happens to a fossil raft of metal rich sediment when it goes down a trench and begins to remelt? Presumably when mixed with brine at gradually rising temperatures and pressures the metal rich sediments are metamorphosed and remelted early, well before the basalts of the old ocean floor. Might they not then become the mineralising solutions of the classical theory and travel back up fissures to the continental surface?

A second question as we saw earlier is again wholly speculative. When an ocean closes and two continents come together, the last fragment of the old ocean may be represented in the ophiolite suites of typical sea floor rocks. When these rocks occur on land, they are often found closely associated with concentration of heavy metal ores. Are these concentrations the remains of brine sinks? If so, the ophiolites and indeed all the possible sites of ocean closure may be the places to look for minerals.

The random discoveries of the oceanographers may therefore have revealed one of the earth's original ore factories, and, coupled with the sea floor spreading they may even result in a powerful predictive tool for land-bound mining geologists. However speculative these notions are, it is quite certain that geologists and oceanographers are only at the beginning of a new line of investigation which may have profound economic significance.

However, back to the practical world. So far as the mining companies are concerned, the known brine sinks have three drawbacks, each of which is fairly crippling. The first is that they are a long way from any market, especially since the Suez Canal become defunct. The second is that they are 6,000 feet down below the surface. The third is that proper assays have still not been done. These drawbacks have not prevented several companies from staking claims. In February 1968, Crawford Marine Specialists Inc., of San Francisco applied to the U.N. for an exclusive mineral exploration lease, but the U.N. rather weakly replied that it had no jurisdiction over an area 50 miles out from any coastline. Another firm, International Geomarine Corporation, adopted a different tactic. It persuaded the Sudanese Government to regard the sinks as lying within its national jurisdiction. A Sudanese company therefore obtained an immediate grant of an exclusive exploration right and the Americans joined in the venture. Next an international consortium concluded that an announcement in an important British newspaper was a fair enough legal title. Presumably regarding the London *Times* as a little too trendy these days it announced its claim in the *Daily Telegraph*, a paper whose banner is still printed in gothic letters.

However exciting these prospects may be, the practical everyday world of mining remains profoundly suspicious of the whole business and its reasons are good ones. There is an almost complete absence of proven hardware capable of operating at depths of 6,000 feet or more. It is almost impossible even to be sure you can repeatedly visit the same place on the bottom, wires from the surface follow a tortuous course because of currents, acoustic signals bend about because of temperature variations, even shore-based radio aids are accurate only to a hundred yards or more. How then can one accurately sample the ore deposit? It is all very well to point out that conditions in the ocean are unlike those on land, but mining companies have learned their expertise the hard way and they have also had a go at ocean mining themselves, they have learned some salutary lessons, even in shallow water where the going is supposed to be easy.

One point must be dealt with, tin dredging in Malaysia is indeed carried out to depths greater than 100 feet in the sea, but not the open sea. When dredging has to be carried out from a ship which moves five feet or more in any direction, the technology required comes out very differently. Everything has to have hinges or universal joints, ships have to be bigger and more seaworthy. Dredgers have a very bad safety record.

Union Corporation (U.K.) Ltd., learned about some of these problems in dredging for tin in St. Ives Bay, Cornwall in 1968. The more easterly part of the bay is exposed to big Atlantic swells and is as good a place for learning about ocean conditions as any. Coastal Prospecting Ltd. had located an area in 36 feet of water or less which contained the heavy mineral cassiterite (tin oxide) derived from the nearby Red River.

The dredger was a converted tanker. It was renamed 'Baymead' and fitted with a suction pipe. The tanks were turned into pump rooms. Navigation was easy because Godrevy lighthouse was nearby and flashing beacons were already sited ashore. Position fixing was therefore dependent on very accurate horizontal sextant angles. The first of many problems came with the vertical motion of the ship in the Atlantic swells which was superimposed on normal rolling and pitching. The effects were coped with by the use of a special drag head on the sea floor and various other buffers and guards on the inboard end of the pipe. The cargo took unexpectedly long to discharge to the shore plant. The combination of quartz and sea water corrosion and 'the wear and tear from operating in rough seas led to almost continuous repairs being required', according to a report presented to the 1969 Commonwealth Mining and Metallurgical Congress in London. Other troubles were due to tidal limitations on the use of Hayle Harbour, and more days were lost from bad weather.

During a good period the operation made a small profit but overall it lost money and was closed down. There was no one single insoluble problem, just a host of little ones which were all overcome in time. It was just that it was all that much more expensive than originally expected.

Another tale of high hopes which slowly subsided like a leaky balloon is that of Sammy Collins' offshore diamonds in South-West Africa. The diamonds occur in extensive marine terraces or raised beaches and have been washed there by rivers which eroded the weathered diamond-bearing igneous rocks farther inland. Mr. S. V. Collins' Marine Diamond Corporation began prospecting in 1961 and sampled the sands under the noses of the giant diamond mining company De Beers. There was a suspicion that the Russians had been doing so too. Collins was successful enough to make De Beers feel it had better join in. De Beers got Ocean Science and Engineering Inc., to fit out a sampling ship, the M.V. *Rockeater*, and formed its own Oceanographic Research Unit. Since 1964 work has gone on steadily exploring the concession at a million dollars a year. Geophysical techniques have been used to map the sea bed thoroughly with the intention of outlining each gravel deposit and revealing such structures as dykes which may trap the heavy diamonds as they are rolled along the beach under the influence of the Benguela current. Reports on the operation are sober reading: 'compared to operations on land, geological information gathered at sea is hard-won. Relatively little information is obtained for a large expenditure of effort, time and money . . .' Apparently direct observation by diving has been as good a technique as anything else. Observation is perhaps the wrong word. Conditions are usually so turbid that the divers cannot see and have to feel their way about, but '. . . surprisingly good results can be obtained in this way and much of the work has been done entirely by feel'. Ocean Science and Engineering took pride from the fact that despite the evident shortcomings of much off-the-shelf oceanographic hardware their sampling equipment came in for high praise. Sampling is by drilling many holes with an annular drill at 10-yard intervals and sucking up the core from inside the drill. One estimate is that a marine diamond deposit would have to be four or five times richer than a land one to be profitable. Here again the enterprise lost money and the only reason De Beers went on was the feeling that the lessons they were learning would be worth the money in the future. Reports of profits in 1970 suggest that the corner may have been turned.

Neither the tin nor the diamond deposits are closely comparable to nodule dredging or brine sink extraction but they have helped to establish a general atmosphere of caution and a feeling that costs proposed by optimists are likely to be much higher in reality especially when the reality involves deep ocean work with round-the-clock transfers from ship to ship of large tonnages of materials.

There are several bits of hardware that probably ought to be developed before anybody tries to mount a full-scale operation. Perhaps the most important is some sort of remote-controlled sea bed chemical analyser device. Toshiba of Japan produced some plans for one at the 1969 Brighton Oceanology meeting. A neutron source bombards the sea bed and the radiation coming back, usually gamma rays, is analysed in an associated spectrograph. An isotope source like Californium 252 produces the neutrons, and a gamma ray detector refrigerated to —100°C measures the characteristics of the stimulated radiation. Such an instrument, however, is potentially as complex as a scientific satellite and the Japanese proposal ducked many of the problems by suggesting that it be hung on an umbilicus from the surface. Nevertheless, a precise measurement of just how much of what metal was on the sea bed would be a powerful source of confidence for those who finally decide to spend the millions required for ocean mining. When people talk of national goals in oceanography and complain that there's nothing quite so clear-cut as launching a satellite or stepping onto the moon they might usefully think of making such a machine. It would open up a great deal of the ocean bed.

15

Farming the Sea

It has been said already that oceanography encompasses many disciplines and will advance on a broad front. No apologies then will be made for swinging abruptly away from the advanced technologies to a much more homely one which may nevertheless result in something of far-reaching importance in the fundamental struggle of getting enough to eat.

The popular impression is that farming the sea or aquaculture is something that laboratory scientists are working on now and which may, in the future, come to be important. In fact there are already hundreds if not thousands of sea farms round the world and you soon discover that visiting them follows a pattern which becomes routine wherever you are.

First you get an address and a telephone number. Usually the place does not appear on the map but you get the impression that it is very remote. When you arrive in the country most people in the government offices in the capital have never heard of their famous fish farmer Dr. X but eventually you get someone to make the difficult telephone call for you since Dr. X has probably not replied to your letter, if indeed he ever got it. At this point it is quite common to discover that the establishment has been moved or at any rate that the important work is being done elsewhere. If you successfully establish contact an exhausting journey follows to some remote terminus where you are met by a practical and much used motor car. Eventually you end up in surroundings of great wildness and beauty where you hear that expansion is just about to begin, that all the problems are within sight of being solved.

The reason for the remoteness of most of the farms that already exist is often the need to avoid pollution. Where there are lots of people and therefore public transport and good postal services there is inevitably pollution. Japan's most famous pioneer Dr. Fujinaga had to leave Takamatsu in the populated island of Shikoku and retreat to the most remote tip of Honshn in Yamaguchi Prefecture. He seems to believe that it is only a matter of time before he is driven out of Japan altogether. Similarly in Britain an exhaustive search of the most remote bays of Western Scotland produced only one good site where access, power supplies, clean water and the absence of pollution made conditions suitable. In North America things are even worse. One of the reasons that aquaculture is so sensitive to pollution is that suitably extensive areas of shallow water are most often found in river estuaries, and the rivers themselves in the so-called advanced nations are frequently little better than diluted streams of sewage.

There are problems of definition with aquaculture. These problems arise from two sources. The first is the attempt to emphasise its similarity with farming livestock on land. By the strict definition true aquaculture would consist of breeding from captive stock, bringing on the livestock on prepared and fertilised pasture and then selling it. In reality nothing is so simple. Sometimes it is not worth hatching from the egg and natural stocks are allowed in and then fattened. Frequently too, fish or other living creatures are fed with food which might itself have been caught from the wild by fishing vessels, or alternatively, they are fed by food grown on land. The second problem arises with the practice of hatching fish and using them to replenish wild stocks which are subsequently caught by normal fishing. An American authority – C. S. Iversen – defines sea farming as 'A means to promote or improve growth, and hence production, of marine and brackish water plants and animals for commercial use by protection and nurture on areas that are *leased and owned*.' The distinction between stocks under private ownership and stocks accessible to anyone seems a fundamental one.

This definition would thus exclude the practice of releasing young fish into bodies of open water in order to supplement commercial fisheries. In any case such attempts have been attended by almost total failure and they are quite likely to cease altogether.

Within the scope of the above definition lie oyster, mussel and other shellfish farms, shrimp and prawn farms and classic fish farming. It would include, with a slight stretch, the selective breeding of salmon to produce superior stocks; although these fish subsequently go out to sea to grow they do return to their original birthplaces to spawn.

However optimistic the forecasts it seems unlikely that sea farming will be of more than marginal significance in the total world food picture. This is especially true if sea farming cannot find ways of functioning economically beyond the salt flats and estuaries. Even if estuary farming is, as some scientists claim, six to ten times more productive than the equivalent area under wheat, there are simply not enough estuarial acres available to be of major significance. The open sea is of course, another matter, there the problems are greater but the promise is greater still.

Even in estuaries the sea farming business is not easy for the pioneers. In Britain the White Fish Authority has put a lot of work into farming fish and its early efforts at Ardtoe can be regarded as a cautionary tale, though the ending may eventually be a happy one.

Ardtoe lies on one of the wild remote peninsulas of Western Scotland which finger out into the Atlantic flanked by long sea lochs. The most westerly part of this particular area is the point of Ardnamurchan which, open to Atlantic swells, is regarded with special alarm by yachtsmen from the Clyde and Oban in the south. To be north of Ardnamurchan in September is to run the risk of not being able to round the Point until the year after, or so the local alarmists say. The Ardnamurchan Peninsula is, except for Land's End itself, the most westerly part of the British mainland and was one of the places where one could be confident of a good clean supply of Atlantic sea water. Most of the books on sea farming begin with a chapter on ensuring the water supply and the British workers were well aware of its prime importance. They chose a beautiful rock-studded bay on the north side of the Ardnamurchan Peninsula where a single track road ran to a very small fishing and crofting settlement called Ardtoe. The idea was to dam off a small acreage by linking some of the rocky islands. Sluice gates would admit sea water as required.

Engineering works were put in hand and in August 1965 the White Fish Authority's scientists set out with a half-million tiny plaice from the Port Erin hatchery in the Isle of Man. They went by sea in a chartered fishing boat and spent the voyage dashing about with oxygen cylinders blowing the life-giving gas into the plastic bags containing the baby fish which were no bigger than small postage stamps. When the boat arrived at Ardtoe there was consternation as it became clear that the engineering work was not really complete. The little fish couldn't stay in the plastic bags much longer so they just had to be put into the sea no matter how many burst bags of cement littered the shore. Surprisingly they seemed to survive the shock of the changed environment quite well in spite of fluctuations in temperature and salinity and the cement bags. But more disaster was in store in the shape of crabs in the sand and mud on the bottom of their new home. Some of the plaice were albinos (hatched plaice often are and no-one seems to know why) so it was possible to see them actually being eaten by the crabs which lay buried in the mud with only eyes and claws protruding. It then began to rain and continued to do so for many days. The low moorland behind Ardtoe drained into the sea farm but pipes had been laid to run off the fresh water into the sea beyond the enclosures. The civil engineers either lacked the imagination to anticipate the rate at which rain can fall in Western Scotland or they just got their sums wrong. The enclosures were flooded with fresh peaty water as the run-off pipes became overloaded. Consequently not only did the acidity of the sea water fluctuate alarmingly but so did the oxygen content because peaty water is particularly low in oxygen.

Another dam was built to pound back the fresh water and the chief scientist, Dr. Shelbourne, ordered that the bottom of the remaining three acres was to be ploughed up and re-sanded. This measure effectively got rid of the crabs. Amazingly some of the infant plaice survived all these natural and human assaults and this fact was a source of some comfort to the scientists.

The idea of free-ranging plaice was dropped in favour of less ambitious investigations into their growth rates and food requirements in netted cages. It was soon discovered that plaice do not mind living on flat sheets of plastic and are quite as happy as those allowed to grub about on the bottom, always provided that they are fed. It was discovered that as many as five fish per square foot could be kept without difficulty. It was clearly important to keep the fish in good condition and feeding in order to combat disease. Survival was found to be better in the floating cages than in the netted cages resting on the bottom. Another interesting discovery was that, like chickens, plaice too have a pecking order. If they were kept in large numbers in a large pool and fed from a single point like the end of a jetty the dominant fish muscled the others out and took all the food.

Although feeding time at the end of the jetty looked good to the visitors the wide disparity in size between the half-starved underlings which could not be seen at the side of the enclosure and the glossy well-fed dominants destroyed the major advantages of fish farming, namely the ability to provide a controlled flow of uniform produce.

To be successful, fish farming does not necessarily have to produce fish more cheaply than the landed price of fish caught in the normal way. Clearly the price cannot be too great but fish farming ought to lead to a reliable supply of fish of the right type and size in a predictably good condition at a known cost. This predictability is what the fish processing industry and the distributing industry wants most of all. Gluts of fish at bargain prices and the occasional complete disappearance of supply are both equally embarrassing.

Although much has been made of fish farming as an undeveloped source of protein for the poor nations of the world most of the promising developments are taking place in rich countries and the species of fish are predominantly luxury fish for the tables of the well-off, who may well eat protein just because it helps to keep them slim. This ironic situation is regrettable but basically unavoidable for both economic and technical reasons.

The economic reasons are obvious enough, even a government-sponsored body like the White Fish Authority in Britain has to show that R and D money is well spent. The prospect of an early profit necessarily implies raising fish which will fetch the maximum price. Plaice is not the ideal fish in this respect but it was chosen because, many years ago, the British Ministry of Agriculture, Fisheries and Food was concerned to augment natural stocks of the fish which supported a large industry. Nowadays interest has moved up market to turbot and Dover sole which are more expensive. Similarly in Japan it is gourmet dishes like the Pacific prawn which have first attracted the entrepreneurs.

Nor is the technology required necessarily suitable for poor nations with large unskilled labour forces. Consider the plant which helps to hatch the plaice in the Isle of Man.

Hatching the eggs and feeding the young fish through their early stages has always been the major difficulty in fish farming. In the wild, mortality at these stages is very great indeed. In 1938 Rollefsen in Norway made a critical discovery when he found that several species of young fish would eat the nauplii of the brine shrimp *Artemia*. The nauplius is the stage immediately after hatching in which the animal is living on the yolk from its egg, is very small, looks like a hairy mite and wriggles about in the water. More important still is the fact that *Artemia* eggs can be easily collected and stored, indeed they must be actually dried before they will hatch. Their natural habitat is the salt pans of inland seas like those near Salt Lake City in the United States.

In the Isle of Man the British applied Rollefsen's discovery to the plaice whose larva, when it has hatched and consumed its own yolk is big enough to eat 10 *Artemia* nauplii a day. To provide the nauplii at the right rate an automated *Artemia* hatching plant was built. The eggs are put into 600 gallon incubator tanks and kept at 23·5°C. for 48 hours. When they hatch debris from the egg cases remains in the vicinity of the nauplii and has to be removed because if the plaice eat this as well as the *Artemia* they are afflicted by digestive upsets. Fortunately the nauplii can be induced to swim away from the egg debris by shining a bright light situated near an automated plug hole. At a pre-set time in the cycle the plug hole opens and shoots the water with the nauplii through a silk screen, the water runs through and the nauplii are collected. They are then automatically dunked into warm water and once more induced to swim away from any remaining egg debris by bright lights. In the last stage they are washed into a concentrator box. This automated nauplii plant works on a repeating cycle and is a most sophisticated piece of systems engineering. One vital bit of it is an electronic counting device which ensures that the number of nauplii presented to the infant plaice are exactly right for their stage of development. As already mentioned, newly hatched plaice eat 10 nauplii a day and those on the point of metamorphosis eat 250 a day. The cost in food of raising a plaice from egg to metamorphosed infant is about a quarter of a new penny each.

Rearing from egg to metamorphosis takes about five weeks and at the end of it the plaice has become a small flat fish living on the bottom, its eyes have moved round and now appear together on one side of the fish, the uppermost side which looks, to layman's eyes, like the fish's back. At this point the fish is weaned on to live red worms which grow on rotting seaweed. This food too, requires to be specially cultured. In a few more weeks the plaice are weaned on to pellets of 'trash' protein with additives bound together in alginate extracted from seaweed. They are then suitable for shipping to Ardtoe for fattening. The White Fish Authority is now confident enough to embark on a demonstration production run

Manipulators cause engineering problems because they have to operate under very high pressure. Having outside hosepipes to transmit hydraulic power can give problems. These 'no-hose' manipulators are from Electric Boat, U.S.A.

Deep Quest from Lockheed is the Rolls-Royce of all the undersea research submarines.

A marine miner with pay dirt. Dr John Mero with manganese nodules.

Riches in view? Manganese nodules on the sea bed at 3,000 feet, off Florida.

A foretaste of marine mining scenes. Work on board Deepsea Ventures' research ship monitoring the picture of the sea bed.

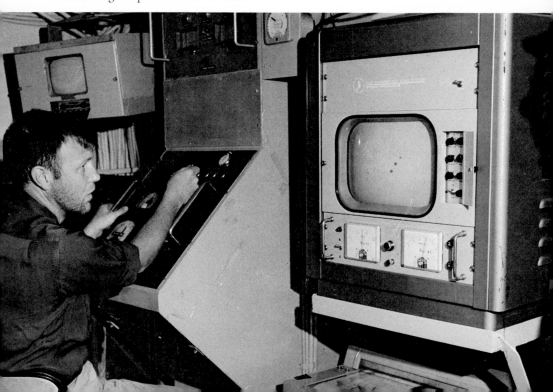

One of the new farmers, Keith Howard, with his best brood stock at Ardtoe.

Japan's successful prawn farm at Yamaguchi.

The 16-foot 'Bumble-
bee' buoy, built by
Scripps Institution of
Oceanography, is lifted
aboard a research ship. A
photopanel camera stores
data from instruments
on film which is collected
every three months by a
servicing vessel. (*Photo:
John D. Isaacs.*)

A monster buoy—
more properly General
Dynamics' Prototype
Ocean Data Station
Bravo—is towed into
position for experi-
mental work in the
north-east Pacific. Sen-
sors for making 100
different measurements
in the air and water are
carried on the mast and
antennae and below the
40-foot-diameter hull.
Data is stored in a 24-
hour memory bank and
transmitted by radio to
an interrogation station
ashore.

which will produce a ton of plaice of marketable size in two and a half years from spring 1970. The equivalent time in the wild to produce fish of similar size would be five years.

If the demonstration is successful the government will be able to set up a pilot producer plant which may convince the food industry that investment in fish farms is worth while.

It will be clear now that there is nothing easy or foolproof about fish farming as it is currently conceived. And so far, the White Fish Authority's experiments have had luck on their side in one important respect, they have been remarkably free from disease. The fish eggs do start off life in an antibiotic dip but apart from that no special measures are taken. Nevertheless a fish vet has been recruited just in case. The possibilities for things to go wrong are very numerous and it can be asserted that it will probably be many years before routines become so well established that farms can be run without highly qualified men. This factor alone means that fish farming is not likely to contribute much to the developing world's food supplies in the near future.

When Ardtoe began the scientists envisaged that one fish per square foot of natural bottom would be a likely maximum density on a free range basis. Now fish at higher densities (more than five fish per square foot) are living in suspended cages and being artificially fed. Not only are the plaice growing faster than anticipated but all sorts of other unexpected things have happened. Very young eels got into the cages and found the regular feeding of the plaice to their liking too. They soon grew too big to escape. Who can say whether this might not be the beginnings of a new profitable line of business? Eels fetch very high prices. Equally unexpected was the behaviour of plaice in the presence of round fish like coley which swim about above the bottom. It had been expected that the plaice would cease to feed when the water was colder than 5°C., however in the presence of coley plaice carry on feeding to lower temperatures. It became clear that it was even an advantage to put in other fish to compete with the plaice, not only did they feed better but, and this is the important point, in terms of fish per unit of capital (the cost of building enclosures) the investment was more productive.

This factor of return on capital is the big unknown in the use of warm water from power stations to bring on fish even faster. At the base load nuclear power station of Hunterston in Scotland, more marine fish farming experiments have been going on in water that is between 6°C. and

10°C. warmer than the ambient sea temperature. Power station cooling water is usually laced with chlorine to stop molluscs and weed growing in the piping, nevertheless plaice and sole were found to live happily enough in water containing up to 0·1 parts per million of free chlorine. Some of them are reaching marketable size in less than two years. The key question turns on the cost of piping the very large flows of cooling water (1,400 gallons per minute) to the fish tanks when they grow to market size in 18 months. Is it worth doing this when it takes only 30 months in the natural sea water at Ardtoe? Another possibility is growing tropical or semi-tropical marine life, like Pacific prawns – will this be economical?

So far the major advantage of being able to hatch fish has not been exploited, namely the ability to breed selectively. The White Fish Authority compares the present position to that of the primitive farmer who did not even know which animals he would be able to domesticate, sheep, cows, goats, pigs or wolves and bears. At the risk of sounding too popular it may be reported that so far turbot seem the most docile and friendly; Mr. Keith Howard who supervises the activity at Ardtoe already sounds slightly like a bloodstock breeder and expresses concern, 'Those are my best two-year-olds, be careful', when visitors balance on the edge of the tanks to photograph the moody flatfish gazing warily up from the bottom.

In terms of the efficiency with which fishy protein may be produced from food it is hard to beat fresh water trout farming with conversion ratios of less than two to one, that is two units of food produce one unit of trout. However, clear streams suitable for trout farming are not common and it costs a good deal of money to ensure that the right conditions are maintained for growth. Trout farms typically have large concrete tanks and complex sluices and other hydraulic works. They also have a steady nagging worry over the danger of drought. Farming the sea will use much cheaper technology of the same class as the fishing industry already uses, that is nets, floats, stiff wire and cable. The aim in farming salt-water fish is to equal the conversion ratios achieved in fresh water and it is widely believed that this will be achieved. If and when it is achieved sea farming will have several other advantages over fresh-water farming. For one thing there will be a far wider range of species and this will help to make fish in general more popular. Furthermore even if a particular fish is not appreciated on the market, the sure and steady supply which farming can guarantee will enable the fish to be promoted as a food. Without such a guarantee marketing campaigns are hardly worth while.

Another advantage of sea water farming stems from the basic laws of physics. In fresh water fish have to expend energy to stay afloat and to keep their body tissues intact. By contrast sea water is denser and has a higher osmotic pressure so salt-water fish get a significant metabolic gain; they are not obliged to expend as much energy just staying alive. Fish have another general advantage too, that of being cold-blooded; they do not have to expend extra energy keeping themselves warm.

Compared to cows, sheep and other land animals, fish are in an even more advantageous position. Cows need to grow big bones to hold themselves upright and consequently they need larger muscles and more energy to move about. If their body tissues were supported in the sea like those of the fish much of the extra energy could go into making edible body weight. One estimate suggests that roughly speaking an acre of land will produce 100 pounds of beef but the same area would produce more than a ton of fish. To emphasise the point, shellfish, which do not move at all, grow even thicker on the ground; from one acre it should be possible to harvest 100 tons of mussels or oysters. American calculations suggest that if Long Island Sound could be given over to mussel fishing its 1,000 square miles could produce a quantity of protein equal to three times the total world fish catch. The Japanese in the Inland Sea already regularly achieve 30 tons of shellfish meats per acre per year.

One of the most efficient marine food producers is the algae chlorella. This organism uses sunlight for highly economic photosynthesis and growth and it has attracted a good deal of attention because it was once tipped as the ideal astronauts' food. In theory it could use sunlight to power a closed circuit of human nutrition during a long space flight. Astronauts would eat chlorella, and more chlorella would grow on their excrement, the only addition to the circuit being energy from sunlight. So far, in spite of its theoretical attractions, it has failed to qualify as economically worth while cultivating on its own, down on the earth, though Japan is reported to have a small production.

The cultivation of molluscs proceeds along traditional lines all over the world. In the richer countries, however, the cost of labour is an increasingly irksome restriction on the expansion of shellfish farming. Mussel cultivation usually depends on harvesting natural seed and relaying the shells at a suitably low density on specially prepared fattening beds. One system relies on rafts from which ropes are suspended. Transplanted mussels are tied on by hand. Evidently to be really attractive in places like the United Kingdom a mechanical system of stripping and re-attaching the seed mussels needs to be developed. Some methods can be very laborious indeed and may involve transplanting and growing colonies of mussels from deep water to closer inshore several times as they mature. Holland produces 48,000 metric tons of mussels a year and France 10,000 metric tons but how long this large production will continue as the non-urban labour force declines in numbers is anybody's guess. Some attempt at mechanisation is being made in the Menai Straits in the beautiful tourist region of North Wales. The initial trials are for oyster seed. The shells are grown on plastic trays stacked on concrete plinths. A fork-lift tractor is used to lift the trays off at low spring tides so that the oysters may be either transplanted or, if all goes well, actually grown to marketable size.

Another area which is benefiting from research aimed at the automation of techniques is hatching the oyster eggs. In Britain the larvae are subsequently reared in glass fibre vats fed with warm filtered sea water to which cultured food has been added together with antibiotics. When the oyster larvae grow out of the free swimming stage after about two weeks they settle on specially prepared surfaces which are slippery so that it is comparatively easy to remove them for relaying on the natural growing beds. The unicellular algal food is prepared separately under the encouragement of light from high-intensity lamps and a steady agitation from air and carbon dioxide injected into the water.

Another important mollusc increasingly neglected because of high labour costs is the clam (*Venus mercenaria*) which grows buried in the mud and is traditionally harvested by a short-handled hoe. It is not as economically suitable for farming as the oyster because it grows more slowly and fetches a lower price at the market. Nevertheless both the Americans and the British have gone to considerable trouble to mechanise the harvesting of the buried shells. The Americans have developed a hydraulic dredge in which a vigorous jet of water washes up the sediment together with the clams buried in it. An escalator lifts the mixture to the surface and on the way up coarse and fine sediment and undersized clams fall through the mesh. The rejected small clams soon bury themselves again and continue to grow.

The British have developed a dredge designed for the cockle (*cardium edule*). The shells are scooped out mechanically and lifted up to the surface by pumping water up a pipe. Again undersized cockles and sediment fall through a mesh back onto the sea bed. The dredge is small enough to be towed by a 35-foot boat. In the first year-long trials which took place in the Thames estuary a commercial return of a ton an hour was gained from beds which were too thin to be harvested by hand. The dredge also harvested beds which were too deep to be dug by hand. The productivity of the system even on a small scale was up to six times better than hand picking. According to the British White Fish Authority which sponsored its development, this system with a little modification will also work for the American clam. Its main limitation is the consistency of the bed in which the cockles are buried, the less like firm sand it is, the worse is the performance of the dredge. Correctly used cockle dredges seem to be ideal farming devices, causing very little damage to the cockles returned to the bed, and scooping up graded shellfish with a productivity that may be up to 20 times the best achieved by hand.

With these newly developed mechanised sea farming techniques both clams and cockles may come back on the market in spite of the disappearance of lowly paid people willing to do the back-breaking work of digging them out of the shore.

Perhaps at this point it is as well to put aquaculture in the perspective of the fishing industry as a whole. Fishing in the open sea produces about 50 to 60 million metric tons a year worth eight billion dollars and all but 10 per cent of the total catch is fish with fins, that is to say that molluscs, shrimps, whales and other marine produce account for only a small proportion of the total. The largest estimate for cultivated marine and brackish water produce is only 10 per cent of the value of the wild catch. It is commonly agreed that the total world fish catch could be increased to around 200 million tons without over-fishing and major economies in the cost of catching the fish could be made with a little international co-operation. According to S. J. Holt, writing in the *Scientific American*, the present value of cod (350 million dollars per year) could be taken with only half the effort currently expended and the money saved (150 million dollars per year) might be used elsewhere to raise the total world catch by five per cent.

That is one way of expressing the fact that mankind is generally too shortsighted to utilise ocean resources on the basis of a maximum sustainable yield. It is usual to hold international conferences whilst the over-fishing proceeds, and when over-fishing stops it is often difficult to tell whether the reason was international agreement or the fact that fishing for a rarer and rarer species eventually becomes uneconomic. The whale has been hunted to the brink of extinction and the blue whale in particular may never recover. The Japanese and the Russians were the last in on that particularly disgraceful chapter but nobody, or rather no nation, ever learns the obvious lessons. Presently Denmark and West Germany, having discovered the North Atlantic migration path of the salmon, are busy fishing that species to death in spite of protests from countries like Ireland, Britain, Norway and Russia which actually produce the salmon. These countries go to considerable trouble to facilitate the passage of the salmon to their fresh-water spawning grounds; in the case of Scotland several hydro-electric plants have installed expensive fish ladders to let the fish pass upstream. When international conferences are called to debate the situation either the nations most at fault, i.e. most efficient at fishing, refuse to attend or obfuscate the issues by ascribing reasons other than over-fishing to the decline in population. More research is usually demanded and scientific quibbles are enlarged to ensure that positive action is not taken because of scientific 'uncertainty'. In the case of the salmon, Denmark claims that the decline is due to the disease ulcerative dermatitis, and that in any case there is insufficient evidence to prove that netting at sea is the cause of a decline in salmon caught in rivers.

This kind of uncompromising behaviour is typical of territorial disputes between nations but what makes matters much more grave at sea is that irreplaceable species are at risk. A brutal question that can be put, and often is put by economists who do not understand biology, is to demand whether it matters or not if species like the blue whale disappear. There are two answers, one is that to take no more than the maximum sustainable yield is sound economics even in the short term, and the second is that if man does some day have to harvest the plankton which form the food of the blue whale, he would probably design a grossly inefficient mechanical blue whale to do so.

It is better planning to have enough blue whales around to keep a stable population. After all they did take millions of years to evolve.

The progress of fishing technology is beginning to accelerate. The picture of the traditional fishermen manoeuvring heavy nets by hand over the side, gutting the fish with a jack knife and returning home after a short trip with the catch preserved in salt or ice is beginning to fade everywhere and not just in the North Sea and around Japan. The use of fish-hunting sonar sets is now common; powered pulley wheels have taken the hard work out of hauling nets; using deep freezing technology to preserve the fish, boats can and do stay at sea longer, using the capital they represent more intensively. In advanced countries a shortage of labour which might have dampened activity is being combated by automation. Gutting fish can now be done by machines which are not necessarily complicated and expensive. Large gutting machines were once limited to big factory ships but smaller cheap ones are now becoming available. In Britain an outstanding mechanical innovator, Mr. James Smith in Shetland, has produced simple reliable machinery which raises on-board processing productivity by five times or more. Within the next two years fish straight from the nets may be tipped into a hopper, sorted for size, aligned, gutted and packed in freezers under the supervision of only one or two men.

These strictly technological innovations mean that over-fishing which used to take say 20 years to run its course may soon show devastating effects in five, even three or two years. Biological innovation has yet to make its impact on the scene but that won't be long now.

In Japan the government's Fishing Agency has been working on acoustical techniques of attracting fish. Hydrophones have been designed to record fish noises with high fidelity and sounds appropriate to various behaviour patterns have been analysed. Carp were used as the experimental subject and their movements were controlled by playing them records of their own bait-eating noises. Japanese biologists next tried the trick with the luxury fish yellow tail which was persuaded to rise from depths of 80 metres so that it could be caught near the surface. Jack mackerel are now the subject of further research along similar lines. Higher power sound projectors will lure fish from greater ranges than the present 500 metres. At the Tokyo laboratories, Dr. Maniwa envisages his sound projectors mounted on buoys in the middle of the fishing fleet. In parallel with this approach is the development of a remote telemetering system which will listen to fish calls and report when suitable concentrations of fish are in its vicinity; only then will it call out the fishing fleet. The Japanese dream of actually replacing the old-style fishing fleet altogether. According to one report when the telemetry buoys have reported to the computer one day in the not too distant future, the computer will

give instructions to each automatically operated fishing boat, that is, a robot ship. Complete with F.M. automatic fish detector, automatic steering devices, automatic net control systems, and various electronic equipment, the boat rushes to the instructed location. There it determines the specific location depth and quantity of the fish involved and automatically selects the most suitable catching method which could be a fully automated drugging (sic) suction pump or an electric fishing which utilizes the field effect of electric current.

Those who are familiar with the heady rush of gadgetry in the land of the economic miracle will not smile at the quaint style, rather they will note Japan is the best placed nation in world fishing and that these dreams are likely to be realised. She has just achieved the biggest fishing industry of all, 11 million tons catch and unlike what was previously the world's largest, Peru's (10 million tons), it is not based on just one fish like the Peruvian anchovy. The Japanese consumer has a highly developed taste for fish which is totally unlike the crude British taste for fried fish and chips. The Japanese eat a wide variety of fish with discernment and one style of eating, sushi, which is raw fish slices on rice, requires cooks who have served several years of apprenticeship. Tempura or prawns fried in batter and another dish called Sashimi now turn up on the menus of good fish restaurants all over the world. Fish has traditionally accounted for most of the Japanese consumption of animal protein fulfilling the role of beef, chicken and pork in other countries. Presently it accounts for two-thirds of the Japanese consumption of animal protein and the demand for the higher priced fish continually rises. Japan's plans to raise total production still see a shortfall of 2·7 million tons by 1977 when the total demand is expected to be 11·6 million tons. The climate for the development of fish farming is thus more encouraging than anywhere else in the world, for this shortfall in supply must be met either by farming or importation and in Japan as we have seen importing is unpatriotic.

Furthermore, as countries like China and India build up a fishing industry, the world's wild stocks are going to suffer further assault or, even if some sort of international co-operation does come about, any one country's share of a limited world stock is likely to be restricted. Either way, fish farming has a sound future, the only problem is the time-scale of development. As already mentioned, a factor common to most of the modern fish farming projects is that they all start with the most expensive fish, and in Japan one pioneer, Prof. Fujinaga, looks forward to riding the wave of prosperity which will give more and more people the money to buy the popular delicacy the Pacific prawn.

A description of his rather unusual enterprise is a convenient way to complete the triptych of aquaculture: the vertebrate fin fish, shellfish like oysters and the crustacea which include crabs, lobsters and prawns.

As always growth rates are the critical factor and this is one reason that lobsters rarely feature in aquaculture. Although they are highly valuable and have no predators they take too long to reach market size. They, like the rest of the crustacea, are scavengers and live on the bottom of the sea sometimes in considerable depths. Some crustacea have an important feature from the point of view of the fisherman—they will stay alive out of the sea for some time because they carry water in their gill cavities. They can therefore be sent to market alive. This is particularly important in the case of the Pacific prawn on the Japanese market.

Prof. Fujinaga first took an interest in the Pacific prawn (*penaeus japonicus*) as a young university graduate 40 years ago. So many older people put him off the idea of working out its life history that he became more than ever determined to do so, in spite of the difficulty. At first he appears to have had no intention of farming it. Studying the prawn became an ineradicable enthusiasm interrupted by the war and afterwards by the task of re-establishing the various national laboratories which had been destroyed. He did 17 years in government service but could not keep his mind off prawns. He managed to pursue his research privately until he knew exactly what prawns eat and in what quantity at every stage of their lives. This is the secret of rearing young prawns from their eggs and unlike young plaice it does not do to give them merely brine shrimp larvae. Fujinaga's procedures have been patented but he's still secretive about details. The diatoms *skeletonema costatum*, *isochrysis* and *monochrysis* feature in the feeding cycle. When the eggs hatch the first stage is again a nauplius, the tiny larva which still carries its yolk. Crustacea, which have their skeletons on the outside, grow by moulting; having shed one skin, the next one to harden is a little larger. Whilst waiting for their new skins to harden prawns are, of course very vulnerable. The nauplius looks more like a spider or mite than a prawn but after six moults it turns into a zoea which is more elongated. After three more moults this becomes a mysis and is big enough to be fed on brine shrimps (*Artemia*). After perhaps a score more moults and now living on the bottom the larva turns into a juvenile, then an adult.

Prof. Fujinaga brings the prawn through the zoea stage by adding fertiliser to the sea water in the hatching tanks. The fertiliser is a mineral one comprising 10 of nitrogen, two of potash and five of phosphate. The appropriate plankton grow on a definite time-scale which matches the development of the prawns. The hatching tanks are 10 metres square and two metres deep, and are continually aerated from pipes at the bottom. The females ready for spawning are either caught in the wild or taken from captive brood stock and about 100 mature prawns are put in each tank. At the right temperature they spawn within one or two nights and plankton cultivation begins.

The acidity of the water and the plankton populations are monitored continuously. The conversion ratio is not particularly good, it takes about 10 pounds of food to produce one pound of prawn but the usual selling price is upwards of six dollars per pound. The total turnover of Fujinaga's company was half a million dollars four years ago and is growing rapidly. In 1970 he was producing 50 tons of high-grade prawns a year. The capital equipment is inexpensive and the scientific procedures are a model of economy. The plankton phases nicely match the food requirements of the growing larvae. Phytoplankton come first, then zooplankton, then the copepods and then the bottom-dwelling organisms follow on. This direct application of fertiliser eliminates all the expense of culturing nutrient plankton in separate tanks yet still gives remarkably high success rates. Each mature female produces up to 500,000 eggs, 40 per cent survive to grow to two centimetres or so and 90 per cent of these grow to the 15 centimetre adult weighing 20 grammes suitable for the tempura market. Fattening takes place in a vast lagoon 200 square kilometres dammed off from the sea and 250 grammes of prawn grow in every square metre of it. Once out of the culture tanks and living on the bottom the prawns are fed on any cheap mussels or baby neck clam. The food is distributed by boat and is simply shovelled over the side on a daily trip.

According to one of Prof. Fujinaga's associates, Dr. Kittaka, 10 million young prawns need the attention of only one technician and a couple of labourers. They work around the year and produce three batches from the lagoon. Fujinaga's establishment is run by a thoughtful and muscular Japanese scientist with commercial acumen called Izumi Yamashita who has a rare and agreeable life. The establishment at the very western end of Honshu is set by a lagoon in an area of low rolling hills with an equable maritime climate. Because pollution and aquaculture never mix, the area is as remote as possible from Japan's industrial anthills. Life has a spacious and uncrowded air but in this isolation no-one has the feeling that the world is passing them by. In fact such is the interest in the cultivation techniques that there is an endless number of people willing to make the journey, bureaucrats from Tokyo, scientists from other parts of the world, entrepreneurs wishing to license the techniques for South-East Asia and journalists wishing to see a sight that will become increasingly familiar in the coming years. Fish farmers all over the world have the rare pleasure of living on the fringes of the industrial whirlpool yet feeling that they are a centre of interest.

16

Understanding Water

BY NOW IT MUST BE CLEAR THAT EXCEPT FOR THE FISHERMEN A SIZEABLE section of the 'ocean business' community regards sea water with no more love than it would lavish on any other large and inescapable obstacle in the path of its ambitions. People concede its value as a medium for getting boats about the world but they could do with a lot less of it, especially vertically. Yet of course the water, the very substance that makes an ocean recognisably an ocean, is itself, to the specialists known as 'physical oceanographers', irresistibly and absorbingly interesting.

It is not particularly easy stuff to study. One of the world's most distinguished research workers in the field of ocean currents, Dr. J. C. Swallow of the National Institute of Oceanography, once began an article on the subject with the flat statement 'There are about three hundred million cubic miles of water in the oceans and all of it is moving about.' Just how it is moving about is a matter of considerable concern for all sorts of practical reasons and there are signs that an increasing amount of effort and money will have to be devoted to finding out.

On the most prosaic level we are concerned with prediction about how waves and tides and currents are likely to affect ships and harbours and beaches. Indeed a major part of the National Institute of Oceanography at Wormley in England was in fact set up as a result of the pressing need of the Allied planners to predict the behaviour of the waves on the beaches of Normandy during the D-day landings of World War II. Since then observations and calculations and experimental work with artificial waves in wave 'flumes' in laboratories have established with some certainty the wave heights that can be expected under particular weather conditions. We even know the statistical likelihood of 'freak waves', terrifying monsters that can occur when big waves in different 'trains' from different sources meet and add their heights together. Once in 100 years we can expect such a wave in the North Atlantic to reach a height of 100 feet though slightly smaller ones could occur as often as once a year. In 1961 the weather ship *Weather Reporter* actually measured a wave 70 feet high and a Russian stereo-photograph exists of a wave which appears to be 82 feet high. These freak waves have probably accounted for some unexplained disasters at sea, especially in the days of sail, and only a few years ago a large modern Italian ship, the *Michelangelo*, was severely damaged by such a wave on her maiden voyage in the Atlantic.

In 1963 wave research took another step forwards, and attracted a good deal of international scientific attention in doing it, when a team from Scripps set out to discover how far waves travel in the open ocean and also how, and how much, their energy is dissipated on the way. It was field work on the grand scale appropriate to a big science and it was conducted over almost the whole span of the Pacific from south to north, a total distance of 10,000 miles.

The sources of waves to be studied were big winter storms down in the high latitudes of the South Pacific and observing stations were set up about 1,000 or 1,500 miles apart in New Zealand, Samoa, Palmyra, Honolulu and Alaska where the waves would end their journey. In the North Pacific where there are no suitable islands the station was the Floating Instrument Platform known as 'Flip' a strange-looking research ship which can be up-ended to float vertically so that she becomes stable enough only to move about three inches in a 30-foot sea.

The wave-measuring instrument specially designed for the project was a very accurate sensor placed on the sea bottom to record the fluctuations in pressure of the water above it – essentially the tungsten wire 'vibratron' that has since proved its value in many other instruments. The signals from the sensor were transmitted by cable and recorded on magnetic tape then put onto punched paper tape for direct input to the computer back at Scripps, far away in California.

As a station picked up the waves arriving from any particular Antarctic storm it measured their period – the time elapsing between the passing of one crest and the next. This was an important measurement because waves travel in groups, the individual leading wave soon dying and being replaced by its successor, the whole group travelling at half the speed of the individual wave. Once the wave period was known the velocity and wave

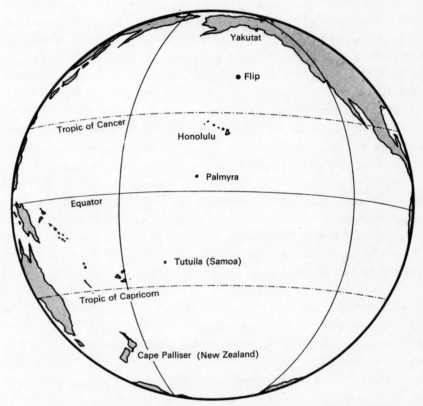

Stations set up by researchers from the Scripps Institution of Oceanography to study waves travelling across the Pacific from New Zealand to Alaska.

lengths of the whole group could easily be calculated and also, of course, the estimated time of its arrival at the next station.

The measurement was simplified because as they advanced the waves gradually spread themselves out, the long period waves travelling faster than the short period ones and so getting farther and farther ahead of them in the lengthening wave train. As they passed station after station it turned out that once they had cleared the confused area around the storm that originated them, perhaps the first 1,000 miles of their journey, the waves travelled the next 9,000 miles with very little alteration or wastage of their energy, far less than had been predicted by calculation. Even near the Equator, where the waves coming in as a swell from the south had to interact with the local waves created by the trade winds, the energy loss was negligible. So it was established that once waves have got themselves organised and left the storm area where they are individually high and where pairs of waves from different trains can cancel each other out, they will carry on almost unimpaired for at least half-way round the world and, if their path was not blocked by land, even farther. (All of which explains, in passing, how waves big enough to gladden the hearts of surfers can arrive on Pacific beaches in an unbroken spell of locally calm weather.)

After all this it is disconcerting to discover that although observations have established adequate rule-of-thumb relationships between wind speeds and wave heights, periods and travel speeds, the understanding of *how* waves acquire their energy from the wind is still not on a properly acceptable mathematical basis. Two theories, each of which in the fifties seemed to have the problem solved, proved to be inadequate. One was that wave generation was directly related to pressure gusts in the air over the sea, the other that the wind coupled into an existing irregular water surface pushing the waves along on their windward side and transferring energy to them. It is now obvious that the dynamics of both the air and water systems are so complex that the final theoretical solution will probably have to wait for far more detailed methods of measuring small-scale atmospheric movements as a basis for calculation.

In their own way, studies of tides, which are just special long-period waves, are in a similar state of disarray. In spite of the authoritative tide-tables available for practically every major harbour in the world the prediction of the behaviour of any particular tide is, strictly speaking, a hit-or-miss affair.

The tides, acting not only on the ocean but on the atmosphere, the earth, buildings, lakes and wells, our own bodies, bathtubs, storage vats, cans of beer or whatever, are caused by the gravitational pull of the moon and the sun. When these bodies are lined up together at new moon and full moon, we get a double dose of pull that causes spring tides with high high waters and low low waters. At quarter and three-quarter moon, when the sun and moon are not in line we get neap tides – low high waters and high low waters.

The relative movements of earth, moon and sun were sorted out physically and mathematically in the late seventeenth century by Isaac Newton and, after 300 years of effort, the astronomical predictions are now very good. So are predictions of the consequential tides raised on the side of the earth farthest away from the direct pull when the earth is distorted or pulled away into the direct tide leaving the water of the distant ocean behind, making secondary tidal bulge which gives many places two tides a day passing over them.

So also, if we want to go into such fine detail, are the variations of the moon's orbit over its 19-year cycle, observed in the eighteenth century by Edmund Halley.

All this is relatively straightforward and presents no problems to mathematicians. In fact Lord Kelvin in the nineteenth century produced a computing machine that works it all out at the turn of a handle. Some of his machines have only recently given way to electronic computers. Yet even though the standard tide-tables are a triumphant example of applied observations, using 'what happened last time' to predict what will happen next, they do not satisfy the aspirations of theoretical physicists. They would like to *understand* the behaviour of the tides as a complete dynamical system, not to produce tide-tables but perhaps to anticipate special effects when special conditions arise, and this at the moment just cannot be done.

The trouble comes from two sources, the *actual* geometry of ocean basins, seas, estuaries and harbours and the *actual* vagaries of the weather. Tides, like any other waves, are affected by their depth over the bottom. As the bottom shoals or grows shallow friction dissipates the tidal energy and the moving water piles up into a higher wave front. Tides are also affected by the natural 'period' of any particular basin in a way that can very easily be illustrated. A coffee cup held in a shaking hand will ripple its contents but not necessarily spill them. If the cup is moved more slowly, at a rate corresponding to its natural period, the coffee will very soon slurp out over the rim. It is the same with larger geographical basins; if their natural period corresponds to that of the tidal movement whether it is a 12-hour high water or 24-hour one, the water will respond in difference of tidal range – in other words the height of its rise and fall. Instead of merely observing effects mathematicians would dearly love to be able to calculate them, at least for 'ideal' oceans and harbours in order to get some better numerical predictions into the real ones.

There have been several recent attempts. One method divides up the important areas into a grid so that each small square can be treated separately according to its true dimensions and offered to a computer for detailed calculation. But the most ambitious calculator in the field is Professor C. L. Pekeris in Israel. He has spent many years on calculating the tides for the whole world oceans on a mathematical model based on the latest and most detailed topographical charts available for each ocean basin. Late in 1969, in a paper forbiddingly entitled 'Solution of Laplace Equations for M_2 tides for the World Oceans' he published the results of some 80 hours of non-stop work by the big GOLEM computer at Israel's Weizmann Institution. One of his findings (or perhaps more correctly the computer's) was an 'amphidromic point' – a position round which the tides apparently rotate but at which there is no measurable tidal stream – in the middle of the South Atlantic. A British research team from the National Institute of Oceanography under Dr. David Cartwright visited the four islands of St. Helena to see whether the result could be checked by observation. As far as they could tell from the limited land-based positions they could use, the computer was right!

If topography causes problems in tide prediction those caused by the weather are even worse, for the obvious reason that it varies from day to day and even from hour to hour. A large area of high pressure literally presses down on the sea whereas a large area of low pressure, the centre of a cyclonic depression, raises the general water level and the height of the tide when it arrives. Apart from this a deep depression and the strong wind gradients around it whip up big waves. Several tidal effects or all of them may combine. A major disaster could be caused by an atmospheric depression raising big storms at the time of a spring tide with unusually high water, in a shoaling basin with a natural period that encourages a big tidal range and strong streams due to the earth's rotational forces. In crowded low-lying inshore areas the loss of life could easily exceed the figure of 2,000 killed in Holland and England by the storm surge of February 1st, 1953, when the tide was eight feet higher than predicted and lasted for several hours longer than the calculated time.

Although this kind of catastrophe threatens many cities, including Venice, it has taken nearly 20 years before London has at last woken up to the danger she runs twice a month in every season of bad weather and is getting down to the expensive business of setting up a movable barrage across the Thames. Statistically the combination of fatal circumstances in the North Sea is likely to occur only once in 400 years but a city of around

10 million people cannot afford to wait and see whether the next storm, still conforming to the average, will be in 800 years time or tomorrow. So once again the study of moving oceanic water leads squarely back to the need for better knowledge of the atmosphere.

Meanwhile there are quite a number of other reasons for investigating the tides. Whereas in the eighteenth century it was necessary to set up national hydrographic offices to save ships from the sea it is now necessary to invest in research on a similar level to save the sea from ships. Leaving aside the pollution hazards of drilling rigs on the continental shelf, the prospect of cargoes of crude oil transported about the world in giant tankers of half a million or even a million tons demands an unremittingly high level of navigation if there are not to be more ecological disasters on a scale that might outclass the *Torrey Canyon* and *Oregon Standard* episodes. It is not just insurance companies and the tourist trade that blanch at these names. When a whole generation of living creatures – fish and seabirds alike – can be wiped out by one oil slick over a whole delicately balanced coastal area there is clearly more at stake than an expensive delay and a red-faced captain when a deep draught ship goes aground. Knowing what the tide is going to do in coastal waters is now as significant as knowing where there are rocks or icebergs.

The science underlying all attempts at finding out more about tides is of course physical oceanography, though this big title probably sounds more appropriate for the newest field of interest in the subject – the study of tides in the open ocean. Until recently almost nothing at all was known about them because the surface of the water provides no stable reference point from which accurate depth measurements can be made. The new observational work is made possible by the new developments in sensitive instrumentation and telemetry.

The advance in technique was led by Walter Munk, the Director of the Institute of Geophysics and Planetary Physics at Scripps. It was he who, with Frank Snodgrass, the engineer who has worked with him on instrumentation projects for over 17 years, (including the celebrated wave-hunt across the Pacific) created the 'tide capsule' with its tungsten wire transducer to descend to the bottom of the ocean and there measure the water passing over it with a sensitivity so acute that it will detect one millimetre in 7,000 metres.

Dumped casually on the deck of a research ship the capsule just might be mistaken for a one-man submarine. Examined more closely it just couldn't. It looks like nothing else on earth – though it might suggest the original 'double bubble' to a frivolous eye mystified by twin inter-connected hollow spheres, each two feet in diameter and constructed of aluminium plate an inch thick to withstand pressures of over 8,000 pounds per square inch. This is the buoyant upper portion of the capsule which is separated from the bottom by 50 feet of cable. In the lower sphere are the recording instruments taping the measurements made by the transducer while the top sphere carried a transmitter-receiver which tells the ship above the state of the instruments and sends out an alert signal if anything goes wrong. This is to enable the ship to make sure that all is well before she leaves the capsule to look after itself for a month or more. A hydro-phone receives instructions from the ship to the capsule, including the command to release itself from the bottom and return to the surface where its tapes of recorded measurements can be collected.

The ultimate research purpose behind this relatively expensive instrumentation is to try to get some measurements of how the tides dissipate the energy, some 2,700,000 megawatts, given to the earth by the gravitational effects of the sun and moon – how, in fact, the tides are slowing down the rotation of the earth and allowing the moon to 'escape' (at the rate of 1/100,000th of its distance from the earth, 2·39 miles, every century so that every century an earthly day grows 1/1000th of a second longer).

Finding the answers to this problem is not likely to be a quick or easy task. Quite probably it will be the final reward for an enormous and widespread effort to study all aspects of the ocean tides, the bonus when the whole picture has been laboriously put together.

On the face of it there are cheaper ways of getting the deep sea tide measurements than developing a foolproof independent whistle-up capsule. The French Hydrographic Service achieved ocean tide measure-ments from 15,000 feet of water by the same vibrotron transducer and tape recording system, without all the additional gadgetry of the Scripps retrieval technique, by simply marking the position of the instrument with a buoy and hauling it up later on a cable. Unhappily (as all oceanographers learn to their cost in wasted experiments) buoys and cables get 'salvaged' by passing fishermen who consider it to be their inalienable right to pick up any 'floating gear' however carefully it is labelled. Cables are attacked and severed by sharks and buoys can break loose from natural causes in bad conditions. A shark-proof buoy capable of riding out a storm is as expensive as a capsule and needs a large expensive ship to lay and weigh it. Also such a buoy and a strong cable stretching to the bottom is a hazard to navigation and if unattended instruments are going to be put on the bottom in any numbers the open sea would become an obstacle course of steel wires. So the whistle-up instrument capsule will probably win the day and be used for all kinds of other observational tasks.

Now that they are sufficiently reliable to sit on the sea bed and record the tides over a complete lunar month the hope is to produce enough capsules to make possible an international survey of deep-sea tides all over the world. Oceanography has been singularly successful in achieving worthwhile results out of such 'co-operative' ventures. The very first time it was tried, in the historic observations of the Transit of Venus in 1769, even with the total lack of anything approaching a modern system of communications and with the accurate measurement of longitude virtually a new discovery, 300 ships of assorted nations took part in the exercise and the transit was triumphantly observed. So during the 1970s there is a possibility that the tide survey will manage to get itself founded by a consortium of nations since, unless the missing information unexpectedly turns up some other way, the lack of it will eventually be felt as a great gap in geophysics.

At the present rate of advance of navigation into ever greater extremes of accuracy, the time will eventually come when it will be necessary to know *exactly* how much allowance has to be made for distortion of the earth's crust by tides and this necessity will be felt by all studies involving a true vertical reference on the earth. Whether expensive research will be undertaken until military or economic pressure forces the pace is, as ever, the test of a big science. Can oceanography continue to command unsecured investment in new knowledge for its own sake and the sake of the path it opens to an unknown destination? The tides, to coin a phrase, will tell.

For the time being the ocean tide research business is not doing too well. The French project has run out of money and is at least temporarily shelved. The Munk-Snodgrass capsules have been developed to a point where they have progressed from testing sorties off California to making successful tide recordings south of Australia but they are not being produced in great numbers. Everyone wants to know more about tides and nobody wants to pay for it. So one of the most hopeful ventures at the time of writing is a British project by the National Institute of Oceanography to make shallow-water recordings of tides arriving from the Atlantic at the edge of the continental shelf.

Along the western coast of the British Isles, section by section, the tides are being measured for a month at a time by rather simple and therefore relatively cheap instruments based on measurement of pressure shown by the varying distance between two plates. They work happily up to a depth of 200 metres with enough sensitivity to detect a variation of water depth of two or three centimetres, and the signals are received by cable to the shore. After the month of measurement is over the instruments are hauled in and moved farther along the coast. In five years or so the whole of the edge of the continental shelf from Norway to Brittany ought to have been covered and the boundary conditions established both for the shallow area and the edge of the deep ocean.

The research is being done by Dr. David Cartwright and so far, apart from routine measurements that will build up the general picture, it has brought to light again the fascinating and unexplained current caused by the diurnal tides off the west coast of Scotland. In effect the tides make an unusual kind of wave that has only a small vertical elevation but a strong flow, enough to suggest at first that the measuring instruments must be picking up some powerful long-period swell from the other side of the ocean. Curiously enough, this odd tidal effect was first reported, without benefit of modern instruments, in 1665 by Sir Robert Morey, one of the founders of the Royal Society. Cartwright discovered his distinguished predecessor when researching the tide literature in the course of publishing his own paper on the subject. Morey had made a field trip to investigate reports of 'Extraordinary Tidal Currents' that ran in the same direction all day between the Hebridean Islands of Harris and Uist, sharing, perhaps, the interest in the subject that later led to Newton's work on the astronomical basis of marine tides.

17

Circulating Systems

KEEPING PACE WITH THE WORK ON WAVES AND TIDES IS THE THIRD BIG
division of the study of oceanic water movements – research on currents.
Of course the real workings of the ocean are a continuous combination of
all three regimes, but any approach we can make to understanding the
whole system depends on first trying to understand each of them
separately.

Most of the great surface currents of the world had been identified and
charted by the mid-nineteenth century – the cold Labrador current that
brings the Arctic icebergs down past the Newfoundland coast, the
Benguela current running up the west coast of Africa, the Humboldt
current taking a similar course up the west coast of South America, the
Agulhas current on the western side of the Indian Ocean, and the Kuroshio
current flowing clockwise across the North Pacific. Some of the hard
work of assembling information about them on charts was done by that
controversial figure in oceanographic history Lieutenant Matthew
Fontaine Maury of the U.S. Navy. By the simple stratagem of offering free
wind and current charts from the U.S. Naval Observatory and Hydrographic Office to the captains of ships which contributed data for their
production Maury collected a prodigious quantity of useful material and
claimed to have shortened by several weeks long sea passages made under

sail – England to Australia from 124 days to 97, Australia to England from
124 days to 63. By 1853 he had been so successful that he was able to
organise a conference at Brussels to engage other maritime nations in the
work of collecting data from their military and merchant ships.

To the physical oceanographers of today the superficial current
pattern is only one part of the whole complex system of currents at every
level that they call the general circulation of the ocean and whose dynamics
they are trying to understand. What is more, they are trying to dis-
entangle this mechanism from the equally complex complementary system
of the atmospheric circulation that partly drives it and is partly driven by
it. Whole books, indeed whole shelves of books, have already been written
on both these enormous subjects and there are plenty more to come. Here
we can only point to the magnitude of the problems.

Put at its simplest, we are dealing with the whole supply of energy
received by the earth from the sun in the form of sunshine, though, as we
have seen in the case of the tides, the gravitational energy is extremely
important as well. Differential solar heating in different latitudes, seasonal
changes of the sun's position, the effects of the earth's rotation and the
effects of the position of alternating surfaces of land and sea are broadly
responsible for the winds which are a transport system for the distribution
of heat. These winds in turn initiate the *superficial* ocean currents. At the
same time the same set of conditions also sets up another system of *deep*
ocean currents driven by variations in the temperature and salinity of the
water from place to place.

Generally speaking the temperature field in the ocean shows a marked
contrast between the warm surface water of the top 600 to 3,000 feet and
the mass of cold water below. There is a sharp division between them,
called the 'main thermocline', where the temperature drops abruptly,
continuing to drop to only just above the freezing point of fresh water
under ordinary sea level pressure. At great depth however, in fact in the
bottom few metres of the ocean, the water temperature actually rises a
little. The reason is unknown and may be either an effect of pressure or
excessive salinity or even sediments in suspension. And when density
differences in water masses are being considered there are those induced by
evaporation of heated surface water in lower latitudes in and near the
tropics, causing differences of salinity. The moving water masses in their
turn transport heat from one latitude to another, and since the quantities of
heat and water vapour entering the atmospheric circulation are critical in

determining its detailed behaviour, it is clear that unravelling the two systems is no simple task.

In the last two decades very great advances have been made in sorting out both the ocean current systems and the broad atmospheric motions, partly because the big computers have arrived to speed the mountainous calculations involved, partly because the growing importance of the subjects have encouraged theoretical work. All of which makes it rather ironic that Matthew Fontaine Maury, over a hundred years ago, was

SURFACE CURRENTS

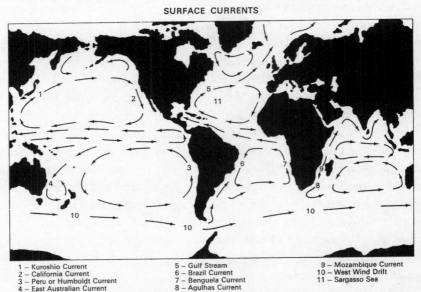

1 – Kuroshio Current	5 – Gulf Stream	9 – Mozambique Current
2 – California Current	6 – Brazil Current	10 – West Wind Drift
3 – Peru or Humboldt Current	7 – Benguela Current	11 – Sargasso Sea
4 – East Australian Current	8 – Agulhas Current	

Principal surface currents of the world make a pattern of circular motions or 'gyres', equatorial counter-currents and strong flows along the western margin of oceans.

busily creating the popular impression that the whole business was pretty well under control.

Fresh from his triumph at Brussels, Maury was encouraged to build a book round the 'Explanations and Sailing Directions' he had written to accompany his famous 'Wind and Current' charts. From the point of view of his final reputation as an oceanographer this was a mistake although the book *The Physical Geography of the Sea*, was a great popular success. Published in 1855 it went into six editions and stayed in print for 20 years until eclipsed by the results of the *Challenger* expedition.

Not being a trained scientist and possessing a strong streak of religious fundamentalism that allowed him to give quotations from Scripture the same weight as careful measurements and observations, Maury got a long way out of his depth in writing about the general circulation. Ignoring both the dynamical problems involved and the serious mathematical work going on to solve them he offered theories of his own, on both oceanic and atmospheric circulations, that infuriated the scientists of the day. A recent editor of the book, Professor Leighly of Berkeley, has even gone so far as to suggest that its major contribution to science was the work of refutation it provoked from Maury's fellow-countryman, William Ferrell. Ferrell, a teacher from Nashville, Tennessee, really started the theoretical work on the motions of fluids on a rotating earth that led to the advances of today.

DEEP CURRENTS

The pattern of the deep oceanic circulation proposed by Henry Stommel repeats the strong flows at the western margins of oceans displayed by the surface movements. Shaded areas denote bottom topography. Black spots in North Atlantic and Weddell Sea show sources of cold water.

Maury and his supporters had maintained that the surface currents were driven by density differences due to unequal solar heating in different latitudes. An opposing view, that they were driven by the drag on the sea of the prevailing winds, had first been put in the eighteenth century by Benjamin Franklin. As has happened so often with rival oceanographic theories, neither was entirely right and neither was entirely wrong. The true explanation was more complex.

As far as we understand the currents today, the superficial system has quite a close correspondence with the pattern of the prevailing winds even though we are still unable to calculate the stress field over the surface from observed winds, as we saw in the study of waves. The result is a pattern

of nearly closed loops called 'gyres' circulating clockwise in the oceans of
the Northern Hemisphere and anti-clockwise in those of the Southern
Hemisphere. Within this general pattern there is also a tendency for
strong currents to lie on the western sides of the oceans.

   To illustrate some of the principles involved in 'decoding' the current
movements it is worth looking in a very general way at the circulation of
the North Atlantic. The prevailing winds – the North-Easterly Trades in
the southern part of the ocean and the Westerlies in the northern part –

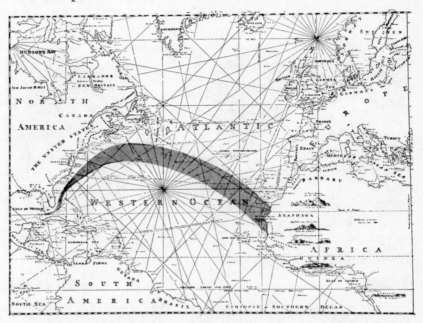

This chart of the Gulf Stream was produced in 1787 as a result of enquiries started by Benjamin Franklin.

give the water a tendency to swirl into a clockwise gyre and at the same
time pile up water against the coasts at the ocean's margins, setting up a
pressure gradient across the current. Meanwhile the northward-moving
water is affected by forces resulting from the rotation of the earth. Any
object on the surface acquires a certain amount of 'spin' from the earth as
they go round together. Spin varies with latitude – a man standing at either
of the earth's poles of rotation is virtually spinning on his own axis while a
man on the Equator is not spinning at all. By changing latitude northwards
a mass of water is moving to a position where it has less spin than the
earth below it. In trying to match its own spin to the spin of the earth, the
water is deflected to the right, reinforcing the existing clockwise motion.

On the eastern side of the gyre, south-going water has more spin than the earth so it is deflected to the left, reducing the strength of the original clockwise flow.

Then there is another related rotational effect to consider – the Coriolis effect – behaving like a force acting at right angles on any body moving on the earth's surface.

In the Northern Hemisphere the deflection is always to the right, in the Southern Hemisphere to the left, and once again the effect varies with latitude from zero at the Equator to a maximum at the poles. Although the

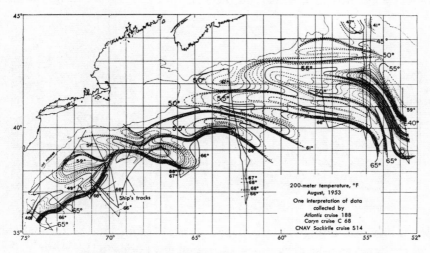

Modern chart of the Gulf Stream made by F. C. Fuglister in 1955.

Coriolis force is always in action, on motor cars, billiard balls, walkers, skiers, indeed on all moving things, it is so small that it can usually be ignored. However, in the case of air or water movements, caused by horizontal forces that are actually extremely small, the Coriolis force becomes significant. In the calculations of currents it must be considered to be acting on every individual particle of water. It not only adds deflection to the wind-driven gyres but it is found 'balancing' the pressure gradients in the water – the tendency to run 'downhill' across the line of the current – so that the flow follows the direction of the isobars, the lines of equal pressure, as it does on the familiar weather maps. In the North Atlantic circulation we find the Coriolis force, among its other effects, 'balancing' the pressure gradient of the warm water heaped up into the equivalent of a hill, several feet high, in the Sargasso Sea on the right-hand side of the

Gulf Stream, which is really a boundary current swinging across the ocean and dividing the warm southern water from the cold water to the north.

The Gulf Stream is not a simple current, in fact the official chartmakers, put to work by Benjamin Franklin in 1770 to plot its position for the benefit of English mail-boat captains, were probably overhasty in taking a mere 17 years to issue their first chart. Two hundred-odd years later no-one would dare to place it so confidently. We know now that it breaks up and re-forms, eddies and meanders, and has narrow streams of very fast water embedded in broader streams that are slower. What is more, when its meanders form loops that are too sharp for the current to follow the

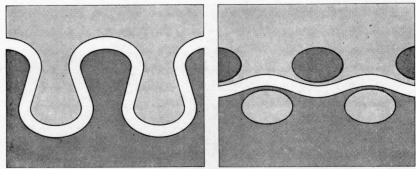

The Gulf Stream acts as a boundary between warm southern water (dark) and cold northern water (pale). When its meanders straighten out anomalous patches of warm or cold water are cut off.

loops break off into great lens-shaped areas of 'anomalous' water that are either hotter or colder than their surroundings. Loops to the north of the stream cut off lenses of warm southern water, loops to the south cut off patches of cold northern water and as these patches can be hundreds of miles across and persist for many months they may have significant effects on the atmosphere and the weather.

Calculating the logic of currents (and countercurrents) in the superficial system of water transport can at least start from a basis of observed winds and drift effects, even from bottles thrown overboard with 'please return' messages in them. But the stress field produced on the surface by the wind does not penetrate much below 300 feet in the sea and up to a point the surface currents in this thin layer of water can be considered as a single transport system.

Calculating the deep currents of the lower transport systems has been much more awkward. Until the last few years it has depended almost entirely on the collection and analysis of samples of water brought up from different parts of the ocean so that differences of temperature and salinity could be plotted over large areas to give an idea of the general density distribution. The basis for a great deal of recent work on deep currents has been the highly productive cruises made in the Atlantic by the German research ship *Meteor* in the 1920s. The meticulous measurements made by her scientists led to an 'atlas' of the properties of the ocean's water at different levels. On the assumption that temperature and salinity measurements would remain roughly the same for long periods, these maps enabled theoretical or mathematical-hydrodynamical models to be worked out for the deep water transport system. Henry Stommel, the most respected oceanographer in this field and a professor at the Massachusetts Institute of Technology, worked out the dynamics of a model of the deep flow in the Atlantic a decade ago. Cold water sinks to the bottom in two areas – the Arctic North Atlantic and the Antarctic Weddell Sea – the Antarctic water being the colder. Everywhere else there is a slow upward movement of water so that the cold water flows slowly from its two sources towards their opposite poles. Stommel predicted the finding of deep western boundary currents formed under the effects of rotation in the same way as the surface currents. He also calculated that the significant link between the wind-driven surface current system and the density-driven deep system occurred in a downward flow around the Equator. Now the tedious processes of sampling, analysis and density plotting are still trying to match up observations and theory.

Recently current measurements have acquired some sophisticated new techniques. Radiocarbon and other types of isotope dating, though difficult to apply, have suggested that the rates at which the deep currents flow are surprisingly slow and that in the Atlantic, for example, a particle of water sinking down from the Weddell Sea will not surface again for around 500 years while in the Pacific it may take as long as 1,500 years. If so the speed of the average deep currents could only be about 1/1,000th of a knot, far slower than anything recorded by direct measurement.

Direct measuring techniques have progressed from tracking a pinger dropped on a parachute to the bottom to the use of the neutrally buoyant float – the 'Swallow' float – designed by Dr. John C. Swallow of the N.I.O. The float is a nine-foot long aluminium tube with a diameter of three inches, weighted so that it will stabilise itself at a predetermined depth or it may be made in the more efficient form of a sphere.

The float contains a pinger and batteries to power it so that it can be tracked for days, weeks, or (in theory at least) even months by a surface ship, tracing out the path of the current over a long distance. The major problem is accurate enough navigation to locate the ship whilst it locates the inch by inch flow of the deep current and there is also the everlasting problem of current studies – variability. Although some major currents may have an all-the-year-round steady flow, floats released in the same place at different times, even only a few weeks apart, can show totally different currents.

Tracking enough floats in enough places to obtain average values is prohibitively expensive in ship-time (and floats themselves are not inexpensive). So now determined efforts are being directed to time-series measurements of currents at different depths by other methods. Where convenient sets of islands allow land stations to be set up in a convenient patch of ocean undistorted by continental shelf effects, sophisticated versions of 'pinging' Swallow floats can be used over a long period and once they have been 'laid' there is no further need for ships. An alternative is to anchor measuring instruments and let them record the direction and strength of current as it passes them. Naturally a single current meter would not reveal what a current did in the next hour or mile after it passed the meter but an array of meters spaced at appropriate distances from each other would monitor a whole area of ocean, or even a whole cube, if meters were set up at different depth levels as well.

On an experimental basis this kind of current measuring has been going on since 1965 at the Woods Hole Oceanographic Institution. Although in terms of ocean coverage this may look like a small-scale experiment involving only a few current meters it is still a major under-taking in terms of design, execution, management, data processing and interpretation and involves quite a number of people and costs around 500,000 dollars a year. The whole programme is run by Dr. Nicholas Fofonoff while Dr. Ferris Webster specialises in the elaborate procedures involved in handling the data.

When the Woods Hole experiment was set up current meters were anchored at various depths below the surface on a year-round basis and their taped recordings were collected at two-monthly intervals for analysis. Although this apparently simple plan ran into severe design problems with corrosion and metal fatigue, the bugbears of all ocean engineering, successful recordings were eventually obtained. The mooring site nearest Woods Hole was the first of a series of five, strung out along

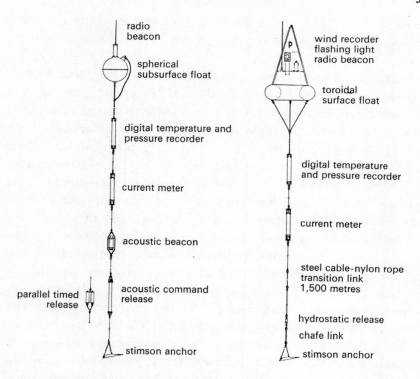

Diagram of instrumentation on two types of buoys with subsurface and surface floats used at Woods Hole for ocean current measurements.

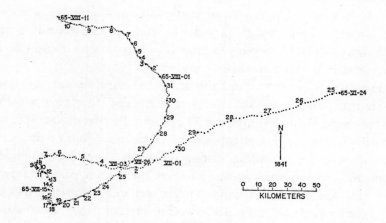

Trace of current motions at a single site between June 24th and August 11th 1965 at a depth of 120 metres.

the 70 °W. meridian on either side of the Gulf Stream over an area where the bottom topography is flat enough not to complicate the current flow.

In general the results have shown that the moored-buoy technique *can* be used to study variability and indeed *should* be used for certain types of measurement. Currents apparently change so much from one short period of time to another that no single set of observations taken, for example, on any one day by one ship, could possibly be regarded as typical at the sites investigated. What also emerges from the recording results is the interesting fact that there are often greater similarities between current flows separated by quite large horizontal distances than there are in flows separated vertically by different water levels at the same site.

If enough time and money are invested in the development of moorings and instruments it should be possible to use these moored buoys to study an even wider spectrum of ocean processes than the experimental recording at Woods Hole can handle at the moment. The Institute of Oceanology of the Russian Academy of Sciences used 20 buoys, each with 15 current meters, in an experiment in a 180 kilometre square of the north-east Atlantic in the first half of 1970. International co-operation could extend such experiments by sharing the cost. Just as arrays of different kinds of seismographs are always at work on land, recording as broad as possible a range of vibrations in the earth, so the current meters could in theory measure every kind of variable flow found in the sea from those that might only occur once a century to microturbulence effects that occur many times a second. Determined mathematical analysis of such data collected over a long enough period and from enough recording sites should also, in theory (especially in key ocean areas) reveal the significant pattern or rhythm of ocean water movements.

It must be said, however, that the sites already in existence have done no more than validate one kind of instrumentation – they are not themselves the magic key to the mystery. The mere collection of data will not necessarily give us an instant explanation of the behaviour of the ocean water any more than measuring earthquake waves led by itself to understanding the cause and significance, let alone the control, of earthquakes. Because this important fact is not understood there is even a danger, pointed out by Professor Stommel himself in a recent paper, that research funds will be poured more enthusiastically into hardware, especially into buoy programmes for routine data collection, whatever the buoys are measuring, than into truly scientific experiments designed to establish what ought to be measured.

'The physical scientist', in Professor Stommel's affronted prose, 'does not see the ocean primarily as a source of wealth or as a jumble of geographic curiosities. He sees it as a hydrographic phenomenon: larger than his laboratory, smaller than a star.' To share that view, to be inspired by the enormous challenge of fundamental research into a gigantic mass of complex material, demands from the finance committee of a government or international agency a sophisticated interpretation of the meaning of science. Such bodies are not given to the grandly simple approach of deciding what they want to know, what they are willing to pay for it and how best to go about it. They are more likely to be seduced by political pressures into underwriting research projects that offer short-term practical results, particularly those with an immediate economic benefit.

Until we know exactly how ocean water moves about we are taking appalling biological risks by interfering with its chemistry. Indestructible toxic chemicals deposited today at the bottom of the Pacific could conceivably return to the surface a thousand years from now to do untold damage. On the other hand the best place to put substances that would gradually decay in salt water might well be deep below the level of the main thermocline where mixing of the top warm layers of water abruptly stops.

Nevertheless one relatively large-scale experiment designed to provide a new approach to the problem of water circulation has got under way with Professor Stommel's approval, indeed it was organised at his original suggestion. This is the GEOSECS Project which will involve taking a large set of ocean water samples for geochemical analysis at 50 different levels from the bottom to the surface. The samples will come from 120 different ship stations strung out along three long traverses running from the Antarctic to the northern boundaries of the Atlantic, Pacific and Indian Oceans. The collection and analysis of the material, providing a complete chemical cross-section of these oceans from top to bottom, will involve two ships for about two years and (excluding ship time) will cost some four million dollars. The most interesting part of the analytical technique will be the interpretation of the movement and age of various radioactive isotopes, whether they occur as the result of natural processes or as the man-made result of bomb fall-out.

The results of this project will undoubtedly shed some new light on some of the biological systems of the sea where at present our information is pitifully inadequate. Even the more straightforward work on currents can help us to find out more about fish, especially the species we want for food. Although we are perfectly familiar with the idea of fishing fleets chasing their prey across the ocean from feeding ground to feeding ground depending on the season, very few people realise how closely the life patterns of most commercially valuable fish are linked to the ocean currents.

If we think of all life in the ocean as a system in which energy from the

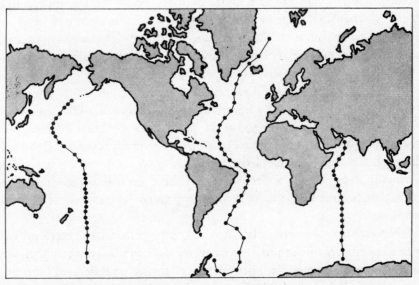

Sampling stations planned for the GEOSECS project to investigate the chemistry of sea water from the surface to the bottom in three oceans.

sun is converted into living flesh we can see how all marine creatures depend on the abundance of the phytoplankton, the 'grass' of the sea, consisting of vast numbers of minute plants, many of them algae as small as a single cell. In the presence of light the algae convert inorganic nutrients from the water into food – the sugars, starches and proteins that enable all animals to live.

The phytoplankton bloom in different latitudes at different seasons and in different quantities – in the spring in the temperate zones – in summer in the high polar latitudes, all the year round in the tropics or in 'upwelling' areas where a steady offshore wind moves the surface water aside and allows nutrients to rise from deeper layers. Where the phytoplankton are

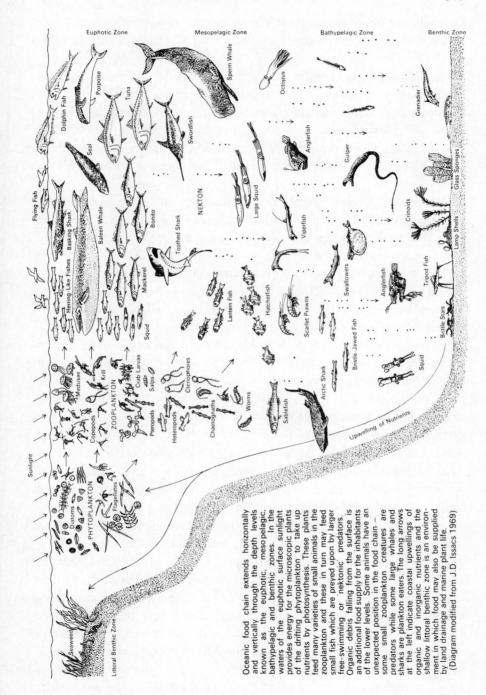

Euphotic Zone Mesopelagic Zone Bathypelagic Zone Benthic Zone

Oceanic food chain extends horizontally and vertically through the depth levels known as the euphotic, mesopelagic, bathypelagic and benthic zones. In the waters of the euphotic surface sunlight provides energy for the microscopic plants of the drifting phytoplankton to take up nutrients by photosynthesis. These plants feed many varieties of small animals in the zooplankton and these in turn may feed small fish which are preyed upon by larger free-swimming or nektonic predators. Organic debris falling from the surface is an additional food supply for the inhabitants of the lower levels. Some animals have an unexpected position in the food chain – some small zooplankton creatures are predators while some large whales and sharks are plankton eaters. The long arrows at the left indicate coastal upwellings of organic and inorganic nutrients and the shallow littoral benthic zone is an environment in which food may also be supplied by land drainage and marine plant life.

(Diagram modified from J.D. Issacs 1969)

there all creatures of the sea must follow, species preying on species in food chains that stretch all the way up to the barracuda, the seal and the killer whale.

Finding the most abundant supplies of food in the sea means for many species a life of migration. Many migrating fish apparently swim upstream in currents to locate and time their spawning so that their developing larvae drift downstream again to nursing areas where the algae will be found in the greatest profusion. The spent fish themselves use currents to return to feeding grounds where they will not compete with or prey on their own young.

The cycles of birth, development to maturity, spawning and recovery may take several years in the life of individual animals and cover hundreds, even thousands of miles, although the pattern is repeated annually by different age groups. We need to know how these cycles work, and the details of the currents and water levels and food concentrations involved before we can tell what factors at what stage in the pattern are, for example, warning signs that stocks are low and fishing should be reduced if the species is not to be endangered.

We also urgently need to know whether or not new agents such as herbicides and pesticides, draining into the sea and carried by currents, can disrupt the balance of phytoplankton and zooplankton abundance and the breeding success of animals all along the food chains. If we do sufficient damage to the chemistry of the ocean it is not beyond us to wipe out certain species at the end of the line. It has already been suggested that sub-lethal doses of pesticides as low as 0·02 parts may interfere with the essential 'imprinting' mechanism which enables a fish to recognise its spawning ground – the familiar area in which it was itself hatched – and without the 'trigger' of this recognition breeding may not take place.

Although it is difficult to put a price on such biological disaster to justify the high cost of research there is at least one area where long experience of trouble has made the figures only too readily available. In every country in the world the official records show how much money could have been saved by more accurate or longer-term weather fore-casting if droughts or floods or hurricanes could have been foreseen or prevented. So it is hardly surprising if the newest branch of circulation research – the study of the ocean-atmosphere 'interface' where so much weather and climate seem to be determined – makes the most immediate appeal to national and international chequebooks. Figures submitted to the

World Meteorological Organisation claim that the current global expenditure on meteorology of 960 million U.S. dollars provides benefits worth *at least* ten times that amount, and that if the present 24-hour forecasts could be extended to even an accurate one-week forecast the benefits could almost double.

Yet when it comes to trying to sort out exactly how the ocean and the atmosphere influence each other's behaviour the complexity and the cost advance together. As it is a physical impossibility to make or use the number of measurements that would be needed to reveal the whole ocean-air relationship in continuous action all over the world, the first thing we need to find out is how fine a network of observation stations we should need to extract any kind of meaningful pattern of measurements. This is the stage of research activity that has been reached at the present time and experiments are now being tried on several levels of scale. Among the latest projects have come the 'monster buoys'.

Monster buoys are really an entirely new kind of unmanned weather station making continuous recordings of large amounts of data in the open ocean and, what is more important for the purposes of weather prediction, making them available by telemetry to receiving stations ashore on demand, normally at least once a day. The new buoys have been developed by General Dynamics for the Office of Naval Research of the U.S. Navy. They have flat pie-dish shaped hulls 40 feet in diameter and seven feet deep with 38-foot high masts supporting antennae and various meteorological instruments. The whole structure, weighing about 100 tons and capable of being anchored in 30,000 feet of water by a cable of two-inch thick nylon line, is at least too large for anyone to steal.

Each buoy can carry 100 different sensors – meteorological ones that monitor the aerial environment and oceanic ones strung out along the mooring line to measure water temperature, pressure, salinity, and sub-surface turbulence at different layers. Everything is recorded on tape so that whether or not short-term data are required for analysis a continuous long-term record is available for long periods. The buoys themselves can operate without human maintenance for a year at a time and carry a two-year supply of fuel for the propane-powered generator systems that provide their electricity.

Obviously the manufacturers of these new stations hope that the two they have built are only the prototypes of a worldwide distribution that would make monster buoys an indispensable part of the whole weather

forecasting network, especially in the huge oceanic areas of the globe
where land-based observatories cannot be located and useful data, apart
from measurements that can be gathered by satellite, are extremely scarce.

While it is undoubtedly true that there is a shortage of weather
stations in the Southern Hemisphere, it is somewhere around this point
that Professor Stommel feels the buoy enthusiasts lose touch with reality.
If a century of forecasting experience has given the meteorologists some
intuitive basis for planning a network suitable for something more refined
than the present one or two-day weather charts, the oceanographers are
not ready to start predicting anything. So here, in particular, the technological tail is trying to wag the scientific dog.

Meanwhile one of the first properly experimental 'jobs for the monster
buoys' has in fact been the investigation of the large areas of 'anomalous'
water found in the North Pacific by a team of oceanographers from
Scripps led by Professor John D. Isaacs. They have been studying sea-
surface temperature and meteorological records made by research vessels
and 'ships of opportunity' for the period 1947–66, and checking the intake
temperatures of water used in ships' boilers. By these means it has been
discovered that patches of water as wide as a thousand miles across can be
as much as eight degrees warmer or five degrees colder than the rest of the
ocean and that these patches persist for long periods of time, certainly for
many months. They are possibly relics of the same kind of process that cuts
off warm or cold 'eddies' to the north or south of the main flow of the
Gulf Stream in the Atlantic. However they are formed they mean the
presence or absence of enormous reservoirs of heat in the ocean area where
a large amount of North American weather is made and they may well be
critically important factors in years of flood or drought. So the monster
buoys have formed focal points in an array of smaller anchored
instrumented buoys set up between Hawaii and Alaska.

On a far larger scale of operation was the BOMEX project of 1969.
This was a one-shot assembly of a big collection of a dozen research ships
and 23 research aircraft with every conceivable kind of instrument and
sensor aboard to saturate with investigations a selected 500 × 500 kilometre
square of ocean surface with the layers of the water systems below it and
the layers of atmosphere above. The project which stands for the 'Barbados
Oceanographic Meteorological Experiment' was the result of recom-
mendations made as long ago as 1962 by the U.S. National Academy of
Sciences. It took years of planning to set up BOMEX and it will be at

least two years of data processing before any worthwhile results emerge from the mass of measurements obtained.

The main objective was to study the ways in which heat energy stored into the upper layers of tropical oceans gets into the lower atmospheric 'boundary' layer – the bottom 5,000 or 6,000 feet – and thence upwards into the other layers of the troposphere, the part of the atmosphere where clouds form signalling the rapid transport of heat from one layer to another.

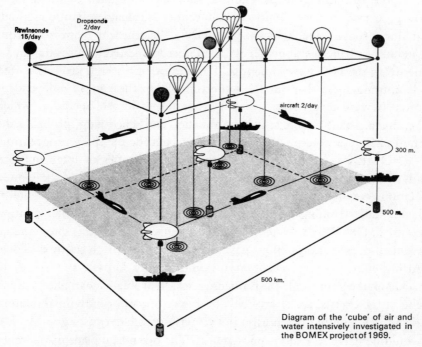

Diagram of the 'cube' of air and water intensively investigated in the BOMEX project of 1969.

These processes are immensely important in the global system of heat balance. Although low tropical latitudes receive more solar heat than the high polar ones the energy emission by the earth as radiation to space is more or less uniform over all latitudes. For this to happen heat must travel from the tropical regions towards the poles and it is the atmosphere, not the oceans, which does the lion's share of the job.

Putting measurements and numbers on the initial stages of this system would be a step nearer to *accurate* long-range forecasts for one or even two weeks at a time. These are not even remotely feasible at present, partly because of the spotty coverage of observation stations which we have already noted, especially in the Southern Hemisphere. There is also another reason. The mathematical models on which computer forecasting depends (and long-range forecasts are essentially a task for computers) can only be devised when the workings of the system are sufficiently well understood for the observational data to be correctly interpreted. As things stand the 'sources and sinks' of atmospheric energy are just not properly worked out. So project BOMEX was designed to supply at least some of the missing factors.

Once this ambitious programme had been announced, to take place over a period of three months 'by all technological and scientific methods at our disposal' which meant all kinds of 'instrument platforms' from satellites to balloons, together with unprecedented data-processing facilities from the U.S. government, all kinds of other research projects started clamouring for a share in the opportunities. In the end 80 independent projects were incorporated as sub-programmes in the operation which became a gigantic exercise in co-ordination and scheduling, probably the most elaborate experiment ever conducted outside the space programme.

BOMEX is likely to be only the first in a series of huge outdoor adventures along similar lines. Not just the United States but the United Nations are anxious to push on as fast as funds will allow to a state of better understanding of atmospheric processes and, if possible, weather control. The World Meteorological Organisation and the International Council of Scientific Unions have now agreed jointly to set up a Global Atmospheric Research Programme (inescapably known as GARP) to tackle on the grand scale the immense research problems that, like so much else in the oceans, are clearly beyond the resources of individual nations.

While it would be churlish not to wish such enterprises good luck it would be naïve to expect quick results. The optimistic plans of solemnly constituted committees conceal the likelihood that the huge energies and dynamic complexities of the planetary systems of air and water circulation will not yield up their secrets without a struggle – a struggle that could cost far more than even an international consortium would at present dare to contemplate. If so we may even be embarking on another task that, like the mapping of the ocean floor, has to be handed on unfinished from one generation to the next.

Epilogue

By the beginning of the seventies oceanography had secured its own future. From the pure science point of view progress had been beyond anybody's wildest dreams. The rise of the theory of sea floor spreading has revolutionised geology and there is now promise of a fundamental understanding of the mechanisms of the deep earth. So here oceanography has a right to stand near or at the head of the queue. Even for other sectors it is clear that support for it no longer depends on the generosity of individual governments. Its development is now firmly linked to forces which are outside their control. In so far as oceanography is a fundamental concern of defence forces, its development is geared to the prevailing degree of international tension. Again because it is now vital to the oil and gas industry it is also geared to the world demand for energy. Whichever way they turn – to the need for fish, weather forecasting, tidal prediction, the need to take a long view of mineral resources, to keep a close eye on pollution – politicians can say that money for oceanography is not just being spent, it is being invested.

Ten years earlier much of the support for oceanography was speculative, a sort of act of faith made at a time when science and technology were high in public esteem. Mr. Harold Wilson won the British General Election on the promise of benefits to come from 'the white heat of the technological revolution'. By the end of the decade technology had almost become a dirty word with science none too far behind, but even if oceanography had been obliged to begin its rapid growth in those unfavourable circumstances, the environmental pollution issue alone would probably still have qualified it for public support.

Indeed in 1969 the director of one of the major American companies on the west coast, whilst admitting that support for oceanography's more optimistic projects was very hard to come by, described pollution as 'a whole new ball park'. His company had made a successful submission to the government on the subject of the starfish *Acanthaster Planci*, known usually as the 'crown of thorns', which was and is eating its way through the coral reefs of the south-west Pacific. Whether this gross disturbance of the natural biological balance was a result of pollution or not is quite unknown. All that is known is that coral had been going strong for at least 400 million years and that suddenly reefs were being wiped out by a plague of starfish to such an extent that people began to speak of 'dooms-day for coral'.

One suggestion is that perhaps the predators of the starfish larvae which have formerly kept the numbers in check may have been affected by the minute traces of D.D.T. now found in the oceans. The traces of D.D.T. are far too small to be directly toxic but maybe 'biological magnification' was taking place. This process is exemplified by the case of the peregrine falcons which proved to have so much D.D.T. and other organochlorines in their bodies that their eggs became too fragile to be viable. They got the organochlorines because they preyed on other birds which themselves had eaten seed dressed with insecticides. The persistent organochlorines had therefore been transmitted along the food chains, getting more concentrated at each stage until they were sufficiently concentrated to kill. This process is now well understood on land but in the sea the food chains are more complex and far less well known. In fact no-one knows whether pollution is the cause of the plague of starfish or not, simply because the detail of what really goes on in the ocean is still more or less a mystery.

The oceans are in a sense the Ultimate Dustbin. Except for solid garbage put into old gravel pits and mine-shafts virtually all the waste from man's activity sooner or later goes back into the sea. Unfortunately the oceans behave unexpectedly towards rubbish and effluents. In spite of all the turbulent overturning we see at the surface, mixing and dilution are far from complete at deeper levels and adjacent bodies of water may remain quite separate from each other as they do, for example, in the lower regions of the Baltic. There the deeper 'stagnant' layers are now becoming very polluted indeed from decades of industrial activity along the shore, and short of pumping out the whole lot into the North Sea it is difficult to see what can be done about it. The boundary between the lower polluted waters and the surface layers is due to sharp changes in temperature and salinity, and therefore density. The water trapped below is de-oxygenated, foul and slowly filling up with sewage.

Putting waste products into the sea is now recognised as being a job that requires great care and far more knowledge than is at present available. The American Academy of Sciences states bluntly that we lack 'the basic scientific background on the physical, chemical and biological processes in the estuarine and coastal marine environments required to treat the problem'. If the sea receives products such as carcasses and organic waste for which it has evolved processing methods over millions of years, there is no problem as long as the quantity it is dealing with is not too great. However, organochlorine insecticides or the steroids in the contraceptive pill which can pass through the human digestive tract and into the sewage are examples of biologically active chemicals which life in the sea has not met before. What happens to them no-one knows because even if they are dispersed the process of biological magnification may reconcentrate them again. The sort of question that now may have to be asked is, for example, what could a slow build-up of contraceptive hormones do to . . . what? shrimps perhaps? The problem is of frightening size because the total number of compounds that the sea had never seen even as late as 1930 is now getting on for 500,000 and there are new ones every year. Of course, there are some limits to the mischief new organic compounds can do. If they are biologically active they are usually broken down eventually, though half lives (when half the substance has been destroyed) may be as long as 10 years. If they are not broken down they are biologically inert and therefore, like polythene or polypropylene plastic, they simply become part of, though an unsightly part of, the sediments and sandbanks. Some compounds have a catalytic function, however, and these may go on working for a long time.

Pollution is an enormous subject and in the context of this book even oceanic pollution can only be dealt with briefly. However, one of the most outstanding points is that many modern pollutants only show up by their effects in the biological environment. Dr. Kenneth Mellanby, Director of the Monks Wood Experimental Station of Nature Conservancy, a well-known authority in this field, admits that often 'we do not know what is happening until a pollutant does some damage.' The living world is, in a sense, mankind's canary capable of giving advance warning of a threat to ourselves. On land in a developed country surveillance of the natural environment ought to work well enough. In Britain for instance there is a vast army of amateur naturalists who instantly report changes in the population or distribution of most wild creatures big enough to be seen with the naked eye. Furthermore, there are enough professional conservationists to follow up the hints. But even this is not good enough. The decline in the peregrine falcon was found quite by chance and related to D.D.T. and other organochlorines at the Nature Conservancy's laboratories only because pigeon fanciers were accusing falcons of catching their birds. These chemicals were found to affect the bird's metabolism of calcium and thus adversely affect the thickness of eggshells. But it is difficult to imagine this feat of biological detection being repeated in the marine environment. Surveillance by large numbers of naturalists does not exist; even if it did, it might not spot something going wrong. We can be sure therefore that pollutants which affect life in the oceans will only show up when they have done relatively far more damage.

All too often such damage as occurs comes as a big surprise. Europe's Conservation Year ended with the withdrawal of millions of tins of tuna fish following the discovery in America that the fish was contaminated with mercury. Previously it had been thought mercury poisoning was a local phenomenon confined to the area around factories discharging mercury-containing effluent. In 1953 the unhappy inhabitants of Minamata Bay in Japan suffered from general destruction of their nervous systems because they ate shellfish gathered near a chemical plant producing P.V.C. plastic. The plant used a catalyst containing mercury. When the American National Institute of Health in Maryland had discovered the cause, restrictions were imposed on gathering shellfish near all similar sources of mercury. But in 1970 it was discovered that mercury was turning up in tuna fish that never came inshore. In fact, tuna fish very sensibly spend their lives as far away from man as possible, circling the centre of the Pacific or the Atlantic Oceans. However, there are no hiding places left and the fish is greatly in demand all over the world. Just how mercury manages to be transmitted from chemical plant effluents (if it really is the case) to the fatty tissues of tuna fish is a mystery. Clearly once again some sort of biological magnification must have been taking place but exactly how is quite unknown. It hardly matters anyway because man's ability to intervene constructively in the food chains of the ocean is negligible. All we can do is stop eating tuna fish for the time being (and no one knows how long that will be) and stop the source of the pollution.

Experience on land behind national boundaries has shown that even when the source of a deleterious pollutant has been traced it is often most difficult to get the discharge stopped. In the sea not only is it difficult to spot the damage early, it is also very difficult to trace the pollutant and when it is traced to say which nation, let alone which company, was responsible. In any case technology is increasingly international; pollution from a French atomic power station is much the same as from a British, Russian or American one.

Tracing the origin of oil slicks can lead to mammoth tangles. Many tankers are owned by nationals of different countries from the country of registration. The crews may be of still different nationality and the oil itself may belong to a fourth nation or even a multinational company. Such was the condition of the *Pacific Glory* which nearly repeated the *Torrey Canyon* disaster by going aground off the Isle of Wight in 1970. Although tanker owners and charterers have got much tougher on prohibiting oily discharges there is still something like a million tons a year of crude oil being discharged into the sea just by tank-cleaning operations by old-fashioned tankers. Modern tankers use the 'load on top' system and keep a spare tank for the washings and water ballast. The oil water mixture has time to separate out in the spare tank so that the oil can be salved, and only comparatively clean water need then be discharged. A million tons of oil may not sound much in an ocean but it has nevertheless led to suggestions that it could 'kill' the Atlantic. This it would do by spreading itself on the surface where, as a layer only one or two molecules thick, it could have a profound and as yet unknown effect on ocean ecology. It might make the planet a better reflector of sunlight and so lower the global temperature, it might inhibit the phytoplankton in their ability to photosynthesise and produce oxygen and therefore help to suffocate us; it might do all sorts of things but the point is nobody knows and no international agreements to improve matters will get anywhere without a bank of unassailable facts that nations will accept as the truth.

Once again we are brought to the conclusion that dealing with the oceans will be a test of man's maturity. The oceans are a complex of

interacting elements which do not recognise national boundaries. Pollution
in one small area eventually affects the whole ocean and both control and
research need to be international. But the oceans offer a challenge to the
nation state and our concepts of national sovereignty in other ways. There
are three other principal areas of concern apart from pollution; one is fishery conservation; the second is military exploitation; and the third is the
question of the ownership of the ocean bed minerals and the control of
ocean engineering. It is perhaps more interesting to consider the last of
these first since the other two are somewhat depressing commentaries on
human carelessness.

At the moment, who owns what on the sea bed is a theoretical problem
since actual exploitation as distinct from exploration has nowhere gone
beyond the present widely agreed boundary of the 'adjacent continental
shelf' which is 600 feet. In fact, 'ownership' is still too strong a word to
describe the attitude of, say, the United States to areas of the sea bed
beyond the traditional three-mile limit of territorial waters. The aim of
nation states is to create a 'stability of expectations' for exploring groups
and companies. If such bodies have confidence that they will be able
reasonably to profit from the fruits of their exploratory labours, they will
then set out in pursuit of that profit with as much willingness as the
present state of technology allows. So far the attitude has been – the shallower the water, the calmer the seas, the nearer the land the better. The
notion that coastal states had some sort of authority over the adjacent sea
bed gained credence after World War II with the United States' unilateral
declaration laying claim to the resources of the sea bed out to the 600-foot
mark. The Geneva Convention of 1958 recognised this sort of claim
adding vaguely that states could go deeper if they had the wherewithal to
do so. The rights allowed the coastal states did not imply occupation or
ownership. The exploration of the North Sea which was divided between
the coastal states came as a direct consequence, the Geneva Convention
having conferred a 'stability of expectations' on the area.

As technology advanced several states pushed the limits of exploration
back. The United States gave one company a phosphate lease in up to
4,000 feet of water near the Forty Mile Bank area. This is only 40 miles
offshore but it is not 'adjacent' since it is separated by a deep trough from
the California mainland. Other states like Australia and Nicaragua have
given exploration permits for similarly deep areas sometimes as far
as 200 miles away. States have been 'expanding their shelves'. Currently

areas up to 200 miles out and up to 6,000 feet deep have been claimed.

There are already problems: what is 'adjacent'? Will it be considered that the shelf around Rockall, the tiny, uninhabited island 400 miles west of Britain is legally part of Britain's continental shelf? It is potentially an oilfield but it is separated from the rest of the continental shelf by a deep trough, so the British have planted the Union Jack on it to make sure. How much of the Blake Plateau off Florida could the British claim on the strength of owning Bermuda and bits of the Bahamas? Supposing the British, who are sceptical of the whole manganese nodule mining concept, put in no claim, how close to British possessions could an American company come – as far as the three-mile limit or the 200-metre contour or what?

Clearly an extension of the national flag land-grab style of settling ownership will not do, or at least we must hope that something more equitable can be devised. Furthermore, many members of the United Nations pertinently ask why should the sea bed riches go to the technologically advanced nations so that once again 'the rich get richer and the poor get poorer'. Malta suggested at the twenty-second General Assembly of the United Nations in 1967 that there should be a freeze on the 'status quo' and that the resources of the deep sea bed should be devoted to the enrichment of mankind as a whole. There are many other pressure groups who advocate that the sea bed should be owned by the United Nations Organisation which might finance itself by leasing the best bits. In this case all the difficulties which ocean exploitation will bring might well be anticipated and reconciled. We saw that if the United States or Japan should decide to open an ocean mine for copper or nickel, it would have a devastating effect on the economy of a developing country that happened to be one of the world sources of the mineral in question. Indeed, one of the reasons for the surge of oil exploration activity in the North Sea and Biscay is the need for European nations to have some weapon with which to resist blackmail by Middle East oil producers. Whether the United Nations would be capable of high-mindedly supervising such goings-on is a question that many observers would answer with a resounding 'No!' The trouble is that the development and conservation of the planetary resources for the good of mankind is not quite the same as the aggregate of the national wishes of the earth's and the United Nations' member states. It ought to be, perhaps, but it manifestly is not.

A study of the problems has been made by William T. Burke for, significantly, the Stockholm International Peace Research Institute. It is called 'Towards a better use of the Ocean'. He points out that 'the supply of certain minerals has been, and is, regarded as critical for national security purposes'. (It has already been mentioned that the U.S. lack of manganese was one of the reasons for interest in manganese nodules in the first place.) There appear to be ample opportunities for mischief and it is quite conceivable that unless, somehow, lines are drawn on the sea bed marking out the areas to be mined as areas to be exclusively occupied by a particular group or power there could well be two rival suction dredges going for the same area. As has already been mentioned, the area of sea bed suitable for mining most cheaply may be quite small until technology develops sufficiently to cope with irregular particle sizes or solid manganese pavements.

In the view of many observers there is no middle ground between the national flag approach, which is how the ownership of land has been settled in the past, and a United Nations style solution. Sometimes 'Intelsat' is mentioned as a solution, an international satellite communications company contracting national expertise as required. Would something along these lines work? Whether such a solution would stand up politically if, for example, it propped up metal prices and restricted supply at a time when Japan, say, really needed something and possessed the technology necessary to exploit the ocean bed unilaterally, well, the answer to that would have to await experience. What is quite plain at present is that decisions on principles need to be taken now before too much is known about ocean resources. The Malta initiative came three or more years ago and so far the signs are that no such steps will be taken. As William Burke says, without a freeze on territorial claims 'the situation could by default tend toward permanent adoption of a flag nation system despite the fact that we are now not sure whether this is the more desirable course to take.' It looks all too likely that the ocean, which really is the planet's last resource, will be consumed, polluted and exploited in broadly the same fashion as the continents have been, with their now threatened burden of wild life.

Perhaps at the mention of wild life it is time to look at the way mankind has exploited the world's fisheries. From a scientific point of view the only sensible thing to do is to arrange that human beings intervene in the food cycles of the sea in such a way that a maximum sustainable yield of desired fish is obtained. The basic mathematics were done in 1931 by E. S.

Russell and have been greatly refined in recent years. There are a number of points which surprise the common-sense view. One is that on a virgin fishing ground the stock may consist of very large old fish in poor condition who successfully get food that young growing fish would like. In such circumstances fishing may actually improve the fishery and give an increased yield by giving the young fish a chance. In the case of a stabilised fishery where the weight of the stock after a season's fishing is equal to the weight at the beginning (that is the total biomass of fish removed is equal to the amount by which it has increased through the growth of little fishes below catchable size), it appears surprisingly enough that the desired stability can be achieved at different levels. In some cases, particularly where large old fish eat all the food, the annual production of a stabilised fishery will be well below its maximum. Usually, however, the reverse is the case and overfishing soon results in falling catches. As a consequence, overfishing means a reduction in the potential yield per unit of effort.

With fish the extinction of a species has rarely, if ever, been achieved because natural populations have a remarkable capacity to recover. Until recently one could be sure that somewhere somehow a few fish would escape but improved technology may make this less true. Travelling on a fast ship like a minesweeper up the North Sea at night it is clear that the nets can extend, with substantial overlaps, for literally hundreds of miles. If you are a fish of catchable size your prospects of escape are dim, you won't get through the nets and getting round them will take you on a very, very long swim.

An individual fish produces millions of eggs, but the same is certainly not true of whales, which, being large mammals with a long period of gestation, reproduce themselves but slowly. In the case of whales not only has overfishing driven the total stock down so far that the industry has collapsed, but, as already mentioned, certain whales, like the biggest of all, the blue whale, have been driven to the edge of extinction. It was thought that so few of these were left that males would be unlikely to be able to get in touch with females for mating. Happily it appears that, using the natural sound channel provided by the thermocline, whales can whistle through the water for perhaps some hundreds of miles, so hope for the revival of the species is not yet extinguished.

The several attempts to impose a quota on the number of whales caught in any one season make melancholy reading. It looks as if there is no known occasion on which any species of fish or marine mammal has been exploited to its maximum sustainable yield by prior agreement. Whales, seals, salmon, yellowfin tuna and sardines have all been driven down in numbers by increasingly competitive overfishing until the species has been nearly exterminated or its continued exploitation become grossly un-economic. Only then have competing nations come to an agreement to limit their activities and there has usually been a good deal of bad grace displayed on all sides. When the parties to agreements include the poorer countries which have come late on the scene and have not amortised their fishing boats, the scientific foundations of the proposed restrictions are often challenged to gain a bit more time. Japan and Russia in the whaling industry were bad examples. Chile, Ecuador and Peru have all clearly been swayed by anti-American political sentiments in their response to measures designed to conserve the yellowfin tuna in the Eastern Pacific.

Many states have regarded the increasing competition in fisheries as a reason for extending their own territorial waters or reserving areas for their own nationals to fish in. Frequently these moves are described as fish conservation measures. Before World War II territorial waters were generally the historical three miles from land (this corresponded to the range of a cannon in previous centuries), when freedom of navigation on the high seas was widely accepted as a sort of human right. By 1951 the U.S.S.R., Colombia and Guatemala had decided that their territorial seas were 12 miles and for purposes of fishing several more states had extended the area of their jurisdiction farther out to sea. Iceland, Canada, and the United Kingdom were among them. There were devious methods of doing so; one was to draw straight lines from headlands and use them as baselines from which the normal range was measured. By 1970 the num-ber of states sticking to a modest three miles was a minority. Most claimed 12 miles, some, notably the Latin Americans, claimed 200 miles.

How much of the advance was due to flag nationalism and how much was due to a desire to conserve fish stocks is hard to say. Either way national jurisdiction over the ocean surface is now being extended in the same way as it is over the ocean bed. Sometimes the two may not be quite as independent of each other as we like to think. In the North Sea trawlers of several nations have taken to trawling near the gas production rigs because fish tend to gather there. Unfortunately the gas pipes leading away to shore along, or just under, the sea bed have become uncovered apparently by the action of trawls. According to one prediction if such a pipe should be ruptured by a trawler a particularly gruesome accident would occur; the escape of high-pressure gas into sea water would produce a vast area of froth which would not support a ship. If the errant trawler itself sank, who would pay compensation to whom, and in what legal framework? Suppose some other ship blundered into the edge of the cauldron, what then? Has Britain any right to stop the accident from happening, or is it not her responsibility at all? If she claims the right to stop the accident from happening then yet another limitation on the right of free passage and free fishing on the high seas will have occurred.

Returning for a moment to the problem of conserving fish stocks it is hard to see how large areas of coastal sanctuary will help since, as we have seen, except for bottom dwellers like plaice, most commercially valuable fish are migratory and may rove several hundred or even several thousand miles. Conservation measures need to be related to what is to be conserved, namely the fish, not to secondary factors like where the fish may go to feed or to spawn. We have mentioned the futility of conserving salmon which spawn in Scottish rivers, when the Danes and West Germans catch them on their Atlantic migrations to West Greenland. Here once again there has been the usual squabble. A conference attended by scientists from 11 countries recommended a 10-year ban on high seas fishing. Denmark, which has no rivers in which the salmon spawn, refused to attend although her interest in salmon is considerable since she takes half the high seas catch. At another conference the Danes quarrelled with most of the scientific evidence which was accepted by 14 out of 18 nations, including even such notoriously unco-operative nations as Russia and Poland. Meanwhile high seas fishing goes on and home-water salmon catches show dramatic declines.

It would be nice if the scientific evidence in fishery conservation could be generated in an impartial organisation but there seems little sign that states will supply enough money for a global fisheries agency. They are fairly reluctant to fund even their own national laboratories and international activity always consumes more money per unit of activity than does the equivalent national one. Meanwhile the economic comedy goes on with most nations deploying at least 20 per cent more effort for their catches than need be deployed if fishing proceeded on a rational basis. Occasionally a whole fishery goes down with a bump; the boats are sold; the bits of equipment go to a museum; the shoreside sheds decay and after a decent interval the whole scene becomes a part of local history. For a particularly eerie and mournful example the connoisseur of fishermen's follies should go to the Shetland Isles where the sunsets linger sadly and long. The local whaling industry brought prosperity in the nineteenth century, until the stocks were totally destroyed, then the Shetlanders went farther and farther from home. Eventually they went south to South Georgia and the Antarctic, and in the end there too the whale stocks were destroyed. Now the whaling is part of history with faded photographs and flensing knives in glass cases. As a man in the museum at Lerwick said, 'Some of Shetland's soul died with it.' It wasn't exactly the Shetlanders fault, everybody had a hand in it: the Norwegians, the Canadians, the Japanese, the Russians, but that's the trouble, it always is someone else's fault until the poetry and the prosperity are only memories.

In no field does the inability of nation states to behave maturely matter as much as in the nuclear weapons field. It may come as a relief therefore to hear that the United States and Russia have endorsed a draft treaty banning nuclear weapons from the sea bed. That does not mean, however, that the sea bed will be demilitarised. Unfortunately, here too original sin is apparent. There is no question that military occupation of the deep sea bed beyond the continental shelf boundaries is technically feasible and may even have occurred. Straws in the wind have been discerned in the remarks of Dr. Robert Frosch, U.S. Assistant Secretary of the Navy for Research and Development at the time (late 1967) of a conference on Law and Security in the ocean. He mentioned underwater silos for nuclear missiles and suggested that being a long way from centres of civilisation (on the Mid-Atlantic Ridge perhaps?) they might be less likely to be the occasion for radioactive contamination. Can defence installations be put beneath the sea bed? The answer may soon be 'yes' according to U.S. oceanography publications. The D.S.R.V. of the U.S. Navy can sub-merge to 5,000 feet (officially) and transfer men at this depth. This depth is within 1,000 feet or so of the top of the Mid-Atlantic Ridge and some of the guyots of the Pacific. Scale models of one atmosphere deep sea habitats have been tested and several U.S. companies like General Dynamics have been issuing glossy brochures: Project Domains is one example: Bottom Fix is another. For the United States the main aim would be to establish a sort of underwater version of the far-flung DEW line, a Distant Early Warning radar chain to monitor Soviet offensive missiles. A submarine DEW line would monitor the passage of Russian missile submarines which it must be noted can already take nuclear warheads to within 12 miles of the Statue of Liberty without offending any of the currently accepted conventions of the sea. One must presume that already sea bed hydrophones for monitoring are installed or ready for installation on the continental shelf of the United States.

One thing is quite clear, no coastal state with the power to stop it would put up with a foreign military emplacement on anything it considers to be its 'own' continental shelf. In such a case 'own' is likely to be regarded as some area over which it has previously claimed jurisdiction either for mineral exploitation or fishery 'protection' or even scientific research. Scientific research has already been limited because the Continental Shelf Convention requires that 'the consent of the coastal state shall be obtained in respect of research concerning the continental shelf and undertaken there.' Little imagination is required to guess at the sort of rigmarole which some states with active bureaucracies might and do require. This uncertainty has then to be fed into the scientists' own uncertain bureaucracy which eventually provides the money to do the work.

It must not be thought that all the scientists' projects are the innocent garnering of a store of knowledge. There are some scientists whose ideas are better characterised as applied science and whose activities could well lead to practical and profound changes in the oceans as we know them. One obvious example is harnessing tides for power generation. The tidal power station at Rance very slightly slows down the earth's rotation. In this example, the changes are predictable but other engineering ideas may be less so. The Russian plan to divert rivers draining from Siberia into the Arctic is a good example, though it is not strictly speaking oceanographic engineering. The argument goes something like this; the Arctic ocean will be deprived of its present supplies of freshwater which tend to stay on the surface of the sea, therefore the sea will not freeze so much since the water at the sea air interface will be saltier and salt water has a lower freezing point than fresh. Therefore the total area of ice and snow will be reduced leading to less sunlight being reflected and more being absorbed. The average temperature of the Arctic will therefore rise and perhaps liberate more water to help raise the total amount of water in the oceans. There are conflicting views on what figures should be put into the equations. The whole effect could be negligible, but it can be readily seen that Russian irrigation plans in southern Siberia could conceivably lead, through their effect on the Polar Ocean, to London, Rotterdam or even Calcutta having to raise their sea walls. No-one knows by what amount the level of the sea might rise nor does anyone know how near the Arctic is to some point at which the changes might be irreversible. The total ice in the Antarctic and Arctic if entirely melted, would raise the sea level by an estimated 300 feet thus drowning most of the civilised world except for Geneva, Madrid, Bogota, Lhasa, and Blaenau Ffestiniog. Happily most of the ice is in the Antarctic and thus safer from civilised interference, but many scientists nevertheless view the whole situation with some degree of alarm.

The story illustrates the way in which almost any consideration of oceanography leads to a systems approach. One of the most important influences of oceanography on the way mankind runs the planet may well lie in this direction. Everything in the ocean itself is not only moving about but interconnected. It is also connected with the land and the atmosphere. In no other study does the almost Swedenborgian complexity of cycles within cycles, systems within systems come across so clearly. It has turned out that the geology of the land is to be comprehended by looking into the sea, and the seas themselves are opened and closed by the land. To test that the Pacific Ocean is closing it will probably be necessary to put a laser on the moon and having made the measurement of the true width of the Pacific and its variations it may become possible to comprehend the geology of the moon by inference from our greater understanding of the geology of the earth. The great slow circulations of the earth's crust influence the quicker circulations of the ocean currents and these the still quicker circulations of the atmosphere. Cutting across these are the biological cycles and the circulations of minerals and sediments.

For man to understand these movements and certainly for man to attempt to intervene in a predictable way, the manner in which he deploys both science and technology has to be systems orientated as well. The implications of a systems approach are too complex to discuss in full but for a good example one may point to the National Aeronautics and Space Administration (NASA) in the United States. Here the central core of policy makers and financiers is surrounded by an inner ring of prime contractors and each of these in turn has another ring of subcontractors, cutting across this concentric arrangement are systems engineers, life support systems, propulsion and guidance systems and so on. On another axis cutting across both are the project men responsible for Mariner, Apollo and so on. Each unit has independence, responsibility and individual drives, but it nevertheless has to continually adjust to the system.

Systems industries in which this pattern is characteristic are likely to be

an increasingly important feature of the rest of the twentieth century. And
that on an international basis is what the great ocean business ought to be
like. Already the trends are plain, the scientists would like investigations of
the air sea interface to be done with a hundred ships all synchronising their
activities in one area. The design of ocean exploration vehicles is already
cross-linked to an array of different requirements. Ocean mining needs to
be linked to world markets and preplanned so that the returns are assured
and the effects are controlled. Conservation needs to bear upon and
inform all national activities in the sea and here, above all, the international approach is the only one which will ensure that the oceans are not
wrecked.

The introduction to this book explained that the aim was to draw
attention to a major new activity of mankind, the aggressive, acquisitive,
technological and scientific mankind now committed to exploring and
occupying the last wilderness and the last planetary resource. No predictions can be made about what will happen. In any case the interpretation of
the trends depends on whether one's view of mankind is optimistic or
pessimistic.

Finally, like Narcissus gazing into the pool, man will look into the
ocean and see only a reflection of himself.

Selected Bibliography

GENERAL BOOKS ON THE OCEANS AND GEOLOGY

Behrman, Daniel, *The New World of the Oceans*. Little, Brown and Company, 1969.

Carson, Rachel, *The Sea Around Us*. Staples Press, 1955.

Deacon, G. E. R. (Ed.), *Oceans*. Paul Hamlyn, 1962.

Engel, Leonard, *The Sea*. Life Nature Library, 1961.

Fraser, Ronald, *The Habitable Earth*. Hodder and Stoughton, 1964.

Gaskell, T. F., *Physics of the Earth*. Thames and Hudson, 1970.

Gaskell, T. F. (Ed.), *The Earth's Mantle*. Academic Press, London, 1967.

Gillespie, Charles Coulston, *Genesis and Geology*. Harvard University Press, 1951.

Hatch, F. H., Wells, A. K. and Wells, M. K., *Petrology of the Igneous Rocks*. Thomas Murby and Co., London, 12th edition, 1961.

Hill, Maurice (Ed.), *The Sea*. 4 vols. John Wiley, 1962-70.

Hills, Sherbon, *Outlines of Structural Geology*. Methuen, London, 1940.

Holmes, Arthur, *Principles of Physical Geology*. Nelson, 1944 and 1970.

Hurley, P. M. (Ed.), *Advances in Earth Science*. M.I.T., 1966.

Keen, M. J., *An Introduction to Marine Geology*. Pergamon, 1968.

King, L. C., *The Morphology of the Earth*. Oliver and Boyd, Edinburgh, 1962.

Kinns, Samuel, *Moses and Geology*. Cassell, London, 1883.

Loftas, Tony, *The Last Resource*. Hamish Hamilton, 1969.

Mavor, James W., Jr., *Voyage to Atlantis*. C. P. Putnam's Sons, 1969.

Menard, H. W., *Marine Geology of the Pacific*. McGraw-Hill Book Company, 1964.

Phinney, Robert A. (Ed.), *The History of the Earth's Crust*. Princeton University Press, 1968.

Scientific American special issue, 'The Ocean', vol. 221, no. 3 (September 1968) – later published in book form.

Shepard, Francis P., *The Earth Beneath the Sea*. Johns Hopkins Press, 2nd edition, 1968.

Shepard, Francis P., *Submarine Geology*. Harper Row, 1963.

Steers, J. A., *The Unstable Earth*. Methuen, 1945.

HISTORY AND DISCOVERY

Books

Challenger Staff Report on the Scientific Results of the Voyage of H.M.S. Challenger During 1873-1876. 32 vols., 1880-95, published by order of H.M. Government.

Darwin, Charles, *Journal of Researches during the Voyage of the Beagle*. Ward Lock and Co., 1839.

Dean, J. R., *Down to the Sea*. Brown, Snow, Ferguson Ltd., 1966.

Mielche, H., *Round the World with Galatea*. William Hodge and Co. Ltd., 1953.

Pettersson, Hans, *Westward Ho with the Albatross*. Macmillan, 1954.

Ritchie, Rear-Admiral G. S., D.S.C., Hydrographer of the Navy, *The Admiralty Chart*. Hollis and Carter, 1967.

Ritchie, Rear-Admiral G. S., D.S.C., *Challenger*. Hollis and Carter, 1957.

Thomson, Sir Charles Wyville, *Voyage of the Challenger*. Macmillan, 1877.

Papers and Articles

Burstyn, Harold L., 'Science and Government in the Nineteenth Century: the Challenger Expedition and its Report.' Paper read at 1er Congrès D'Histoire d'Oceanographie, Monaco, 1966.

Meyerhoff, A. A., 'Arthur Holmes, Originator of Spreading Ocean Floor Hypothesis.' *Journal of Geophysical Research*, vol. 73, no. 20 (15 October, 1968), pp. 6563-9.

Ritchie, G. S., 'The Royal Navy's Contribution to Oceanography in the

XIXth Century.' Paper read at 1ᵉʳ Congrès D'Histoire d'Oceanographie, Monaco, 1966.

Rupke, N. A., 'Continental Drift before 1900.' *Nature*, vol. 227 (25 July, 1970), pp. 349-50.

CONTINENTAL DRIFT, SEA FLOOR SPREADING AND PLATE TECTONICS

Books

Bascom, Willard, *A Hole in the Bottom of the Sea*. Doubleday and Company Inc., New York, 1961.

du Toit, A. L., *Our Wandering Continents*. Oliver and Boyd, 1937.

Gutenberg, B. and Richter, C. F., *Seismicity of the Earth, and Associated Phenomena*. Princeton University Press, 1949.

Irving, E., *Palaeomagnetism and its application to Geological and Geophysical Problems*. Wiley, 1964.

Runcorn, S. K. (Ed.), *Continental Drift*. International Geophysics Series, vol 3, Academic Press, 1962.

Runcorn, S. K. (Ed.), *International Dictionary of Geophysics*. Pergamon Press, London, 1968.

Suess, E., *Das Antlitz der Erde*. F. Tempsky, Vienna, 1888. Translated by H. B. C. Sollas as *The Face of the Earth*. Clarendon Press, Oxford, 1906.

Tarling, D. H. and M. P., *Continenal Drift*. G. Bell, 1971.

Wegener, A. *The Origin of Continents and Oceans*. First English Edition, Methuen, 1924, translated from 3rd (1922) German Edition.

Papers and Articles

Bullard, E. C., 'Reversals of the Earth's Magnetic Field'. Bakerian Lecture, 1967. *Philosophical Transactions of the Royal Society of London*, Series A, *Mathematical and Physical Sciences*, vol. 263, no. 1143 (12 December 1968), pp. 481-524.

Bullard, E. C., Everett, J. E. and Smith, A. G., 'The fit of the continents around the Atlantic'. *Phil. Trans.*, Series A, vol. 258 (1965), pp. 41-51.

Bullard, E. C. and Gellman, H., 'Homogeneous Dynamos and Terrestrial Magnetism'. *Phil. Trans.*, A, vol. 247 (1945), pp. 231-78.

Bullard, E. C., Maxwell, A. E. and Revelle, R. R., 'Heat Flow through the Deep Sea Floor'. *Advances in Geophysics*, vol. 2 (1956), pp. 153-81.

Carey, S. W., 'Continental Drift: a Symposium'. University of Tasmania, Hobart, 1958.

Cox, A., Dalrymple, G. B. and Doell, R. B., 'Reversals of the Earth's Magnetic Field'. *Scientific American*, vol. 216 (1967), pp. 44-54.

Cox, A., Doell, R. B. and Dalrymple, G. B., 'Geomagnetic Polarity Epochs and Pleistocene Geochronology'. *Nature*, vol. 198 (1963), pp. 1049-51, and other papers.

Crawford, Arthur Raymond, 'Continental Drift and Un-Continental Thinking'. *Economic Geology*, vol. 65(1970), pp. 11-16.

Degens, Egon T. and Ross, David A., 'The Red Sea Hot Brines'. *Scientific American*, vol. 222, no. 4 (April 1970).

Dewey, J. F., 'Continental Margins: a Model for Conversion of Atlantic Type to Andean Type'. *Earth and Planetary Science Letters*, vol. 6 (1969), pp. 189-97.

Dewey, J. F., 'Evolution of the Appalachian-Caledonian Orogen'. *Nature*, vol. 222, no. 2189 (12 April 1969), pp. 124-9.

Dewey, J. F., and Horsfield, B., 'Plate Tectonics, Orogeny and Continental Growth'. *Nature*, vol. 225 (February 1970), pp. 521-5.

Dickson, G. O., Pitman, W. C., III, and Heirtzler, J. R., 'Magnetic Anomalies in the South Atlantic and Ocean Floor Spreading'. *Journal of Geophysical Research*, vol. 73, no. 6 (15 March 1968).

Dietz, Robert S., 'Continent and Ocean Basin Evolution by Spreading of the Sea Floor'. *Nature*, vol. 190, no. 4779 (3 June 1961), pp. 854-7.

Dietz, Robert S. and Holden, John C., 'The Breaking of Pangaea'. *Scientific American*, vol. 223, no. 4 (October 1970).

Elsasser, W. M., 'Interpretation of Heat Flow Equality'. *Journal of Geophysical Research*, vol. 72 (1967), pp. 4768-70.

Garland, G. D. (Ed.), 'Continental Drift'. Royal Soc. Canada Special pub. no. 9 (Ottawa, 1966), pp. 1-140.

Gaskell, T. F. and Swallow, J. C., 'Seismic Experiments on two Pacific Atolls'. Occasional Papers of Challenger Society, 1953.

Girdler, R. W., 'The Role of Translational and Rotational Movements in the formation of the Red Sea and Gulf of Aden'. In *The World Rift System*, Geol. Surv. Can. Paper 66-14 (1967), pp. 65-77.

Glass, B. P. and Heezen, B. C., 'Tektites and Geomagnetic Reversals'. *Nature*, London, vol. 214 (1967), pp. 372-4 (also *Scientific American*, vol. 217, pp. 32-38).

Hayes, D. E. and Pitman, W. C., 'Magnetic Lineations in the North Pacific'. (Preprint circulated July 1969).

Heezen, Bruce C., 'The Rift in the Ocean Floor'. *Scientific American*, vol. 203, no. 4 (October 1960).

Heezen, B. C., Tharp, M. and Ewing, M., 'The Floors of the Oceans, 1: The North Atlantic'. Geological Society of America, Special Paper 65 (1959).

Heirtzler, J. R., 'Sea Floor Spreading'. *Scientific American*, vol. 219, no. 6 (December 1968).

Heirtzler, J. R. and Le Pichon, X., 'Crustal Structure of the Mid-ocean Ridges. 3: Magnetic Anomalies over the Mid-Atlantic Ridge'. *Journal of Geophysical Research*, vol. 70 (1965), pp. 4013-33.

Heirtzler, J. R., Le Pichon, X. and Baron J. G., 'Magnetic Anomalies over the Reykjanes Ridge'. *Deep Sea Research*, vol. 13 (1966), pp. 427-43.

Heirtzler, J. R., Dickson, G. O., Herron, E. M., Pitman, W. C. and Le Pichon, X., 'Marine Magnetic Anomalies, Geomagnetic Field Reversals and Motions of the Ocean Floor and Continents'. *Journal of Geophysical Research*, vol. 73 (1968), pp. 2119-36.

Hess, H. H., 'Drowned Ancient Islands of the Pacific Basin'. Paper to the American Geophysical Union, Washington D.C., 27 May 1946.

Hess, H. H., 'History of Ocean Basins'. *Petrological Studies:* a volume in honour of A. F. Buddington (November 1962), pp. 599-620.

Hess, H. H., 'Mid-Oceanic Ridges and Tectonics of the Sea Floor'. Paper circulated as preprints, December 1960, presented at National Academy of Sciences symposium, April 1961, and used in expanded form as a Presidential address to the Geological Society of America, November 1963.

Hess, H. H., 'Mid-Oceanic Ridges and Tectonics of the Sea Floor'. *Submarine Geology and Geophysics*, ed. W. F. Whittard and R. Bradshaw (Butterworths, London), pp. 317-33.

Hess, H. H. and Buell, M. W., Jr., 'The Greatest Depth in the Oceans'. *Transactions of the American Geophysical Union*, vol. 31, no. 3 (June 1950), pp. 401-5.

Hill, Maurice N., 'The Topography of the Mid-Atlantic Ridge'. Paper read at Meeting of Association d'Oceanographie Physique, Rome, 1954.

Horsfield, B. and Dewey, J. F., 'How Continents are Made and Moved'. *Science Journal*, vol. 7, no. 1 (January 1971).

Hurley, P. M., 'The Confirmation of Continental Drift'. *Scientific American*, vol. 218, no. 4 (April 1968).

Hurley, Patrick M. and Rand, John R., 'Pre-Drift Continental Nuclei'. *Science*, vol. 164, no. 3885 (13 June 1969), pp. 1229-42.

Isacks, B., Oliver, J. and Sykes, Lynn R., 'Seismology and the New Global Tectonics'. *Journal of Geophysical Research*, vol. 73, no. 18 (15 September 1968), pp. 5855-99.

McKenzie, D. P. 'Plate Tectonics of the Mediterranean Region'. *Nature*, vol. 226 (18 April 1970), pp. 239-43.

McKenzie, D. P. and Morgan, W. J., 'The Evolution of Triple Junctions'. *Nature*, vol. 224 (11 October 1969), p. 125.

McKenzie, D. P. and Parker, R. L., 'The North Pacific: an Example of Tectonics on a Sphere'. *Nature*, London, vol. 216 (1967), pp. 1276-80.

Mason, R. G. and Raff, A. D., 'A Magnetic Survey off the West Coast of North America, 32°N to 42°N'. *Bulletin of the Geological Society of America*, vol. 72 (1961), pp. 1259-65, and other papers.

Matthews, D. H., 'Mid-Ocean Ridges'. Entry in *International Dictionary of Geophysics*, ed. S. K. Runcorn (Pergamon Press, London, 1968).

Menard, H. W., 'The Deep Ocean Floor'. *Scientific American*, vol. 221, no. 3 (September 1969), pp. 126-142.

Menard, H. W. and Atwater, Tanya, 'Changes in Direction of Sea Floor Spreading'. *Nature*, vol. 219 (3 August 1968), pp. 463-7.

Morgan, W. Jason, 'Rises, Trenches, Great Faults and Crustal Blocks'. *Journal of Geophysical Research*, vol. 73, no. 6 (15 March 1968).

Morley, L. W. and Larochelle, A., 'Palaeomagnetism as a Means of Dating Geological Events'. Royal Society of Canada, Special Pub. 8 (University of Toronto Press, 1964), pp. 39-54.

Opdyke, N. D. and Foster, J. H., 'Reversals of the Earth's Magnetic Field'. *Science Journal* (September 1967), pp. 56-61.

Opdyke, N. D., Glass, B., Hays, J. D. and Foster, J., 'Palaeomagnetic Study of the Antarctic Deep-Sea Cores'. *Science*, New York, vol. 154 (1966), pp. 349-57.

Le Pichon, Xavier, 'Sea-Floor Spreading and Continental Drift'. *Journal of Geophysical Research*, vol. 73, no. 12 (15 June 1968), pp. 3661-97.

Pitman, W. C., III, 'Sea Floor Spreading'. *Science Journal* (February 1969), pp. 51-56.

Raff, A. D., 'The Magnetism of the Ocean Floor'. *Scientific American*, vol. 205, no. 4 (October 1961).

Rikitake, T., 'Oscillations of a System of Disc Dynamos'. *Proc. Camb. Phil. Soc.*, vol 54 (1958), pp. 89-105.

Roberts, D. G., Bishop, D. G., Laughton, A. S., Ziolkowski, A. M., Scrutton, R. A. and Matthews, D. H. 'New Sedimentary Basin on Rockall Plateau'. *Nature*, vol. 225 (10 January 1970), pp. 170-2.

Rothé, J. P., '*La Zone Seismique Mediane Indo-Atlantique*'. Paper at discussion meeting, 1954.

Runcorn, S. K., 'Towards a Theory of Continental Drift'. *Nature*, vol. 193 (1962), pp. 311-14.

Smith, A. Gilbert and Hallam, A., 'The Fit of the Southern Continents'. *Nature*, vol. 225, (10 January 1970), pp. 139-44.

Stride, A. H., 'Mapping the Ocean Floor'. *Science Journal*, vol. 6, no. 12 (December 1970).

Suffolk, G. C. J. 'Precession in a Disc Dynamo of the Earth's Dipole Field'. *Nature*, vol. 226 (16 May 1970), pp. 628-9.

Sykes, Lynn R., 'Mechanism of Earthquakes and Nature of Faulting on the Mid-Ocean Ridges'. *Journal of Geophysical Research*, vol. 72, no. 8 (5 April, 1967).

Sylvester-Bradley, P. C., 'Tethys: the Lost Ocean'. *Science Journal*, (September 1968).

Symposium on the Red Sea. Royal Society, 29 October 1970, Series A, no. 1181.

Tams, E., '*Die Seismizität du Ozeane und Kontinente*'. *Zeibschrift für Geophysik*, 4 Heft. 5 (1928), pp. 245-6, 4 Heft. 7-8, pp. 321-48.

Tazieff, H., 'Tectonics of the Northern Afar (or Danikil) Rift'. *Upper Mantle Project Scientific Reports*, no. 27 (1970).

Tazieff, H., 'Exposed Guyot from the Afar Rift, Ethiopia'. *Science*, vol. 168 (29 May 1970), pp. 1087-9.

Tazieff, H., 'The Afar Triangle'. *Scientific American*, vol. 222, no. 2 (February 1970).

Vacquier, V., Raff, A. D. and Warren, R. E., 'Horizontal Displacements in the Floor of the Pacific Ocean'. *Bulletin of the Geological Society of America*, vol. 72 (1961), pp. 1251-358.

Vali, Victor, 'Measuring Earth Strains by Laser'. *Scientific American*, vol. 221, no. 6 (December 1969).

Vening Meinesz, F. A., 'Convection Currents in the Earth and the Origin of the Continents'. *Proc. K. Ned. Akad. Wet.*, B, vol. 55 (1952), pp. 427-553.

Vening Meinesz, F. A., 'The Earth's Crust and Mantle'. *Developments in Solid Earth Geophysics*, vol. 1 (Elsevier, Amsterdam, 1964), pp. 1-124.

Vine, F. J. and Matthews, D. H., 'Magnetic Anomalies over Ocean Ridges'. *Nature*, vol. 199 (1963), pp. 947-9.

Vine, F. J. and Wilson, J. T., 'Magnetic Anomalies over a Young Ocean Ridge off Vancouver Island'. *Science*, New York, vol. 150 (1965), pp. 485-9.

Wilson, J. T., 'Evidence from Islands on the spreading of Ocean Floors'. *Nature*, vol. 197 (1963), pp. 536-8.

Wilson, J. T., 'Continental Drift'. *Scientific American*, vol. 208, no. 4 (April 1963).

Wilson, J. T., 'A New Class of Faults and their bearing on Continental Drift'. *Nature*, London, vol. 207 (1965), pp. 343-7.

Wilson, J. T., 'Transform Faults, Oceanic Ridges and Magnetic Anomalies southwest of Vancouver Island'. *Science*, New York, vol. 150 (1965), pp. 482-5.

BIOLOGY, ZOOGEOGRAPHY AND CHEMISTRY

Books

Calvin, Melvin, *Chemical Evolution*. Oxford University Press, 1969.

Darwin, Charles, R., *The Origin of Species by Means of Natural Selection*. John Murray, 1859.

Harden Jones, F. R., *Fish Migrations*. Edward Arnold, 1968.

Hardy, Alister C., *The Open Sea*, Part 1: The World of Plankton, Part 2: Fish and Fisheries. William Collins, 1959.

Iverson, E. S., *Farming the Edge of the Sea*. Fishing News, London, 1968.

Leach, Gerald, *The Biocrats*. Jonathan Cape, 1970.

Marx, Wesley, *The Frail Ocean*. Ballantine Books, 1967.

Papers and Articles

Berggren, W. A., 'Rates of Evolution in some Cenozoic Planktonic Foraminifera'. Preprint 1969.

Charig, A. J., 'Kurtén's Theory of Ordinal Variety and the Number of the Continents'. In the press.

Cox, C. Barry, 'Migrating Marsupials and Drifting Continents'. *Nature*, vol. 226 (23 May 1970), pp. 767-70.

Ericson, D. B., Ewing, M. and Wollin, G., 'The Pleistocene epoch in deep-sea sediments'. *Science*, vol. 146 (1964), pp. 723-32.

Fox, Sidney, 'In the Beginning – Life Assembled Itself'. *New Scientist*, vol. 41, no. 638 (27 February 1969).

Funnell, B. M., 'Pre-Quaternary Occurrences of Micro-fossils in the Oceans'. *Micropalaeontology of Oceans*, SCOR Symposium Volume (Cambridge University Press, 1968).

Funnell, B. M. and Smith, A. G., 'Opening of the Atlantic'. *Nature*, vol. 219 (1968), pp. 1328-33.

Harrison, C. G. A. and Funnell, B. M., 'Relationship of Palaeomagnetic Reversals and Micro-palaeontology in two Late Caenozoic Cores from the Pacific Ocean'. *Nature*, London, vol. 204 (1964), p. 566.

Hays, James D. and Opdyke, Neil D., 'Antarctic Radiolaria, Magnetic Reversals and Climatic Change'. *Science*, vol. 158, no. 3804 (24 November 1967), pp. 1001-11.

Isaacs, John D., 'The Nature of Oceanic Life'. *Scientific American*, vol. 221, no. 3 (September 1969), pp. 146-67.

Kennet J. P. and Watkins, N. D., 'Geomagnetic Polarity Change, Volcanic Maxima and Faunal Extinction in the South Pacific'. *Nature*, vol. 227 (29 August 1970).

Kurtén, Björn, 'Continental Drift and the Palaeography of Reptiles and Mammals'. *Commentationes Biologique Societas Scientiarum Fenkica*, vol. 31, no. 1 (1967), pp. 1-8.

Kurtén, Björn, 'Continental Drift and Evolution'. *Scientific American*, vol. 220, no. 3 (March 1969).

Martin, P. G., 'The Darwin Rise Hypothesis of the Biogeographical Dispersal of Marsupials'. *Nature*, vol. 225 (10 January 1970), pp. 197-8.

Preliminary Proposal for the Geochemical Ocean Section Study, January 1969.

Scientific American – 'The Biosphere' (special issue), vol. 223, no. 3 (September 1970).

THE MOON AND SOLAR SYSTEM

Books
Glasstone, Samuel, *The Book of Mars*. N.A.S.A., 1969.
Urey, H. C., *The Planets*. Yale, 1952.

Papers and Articles
Hatherton, Trevor, 'Lunar Continental Migration and Maria Spreading'. *Nature*, vol. 225 (28 February 1970), pp. 844-5.

Kane, M. F., 'Doppler Gravity: a New Method'. *Journal of Geophysical Research*, vol. 74 (1969), p. 6579.

'The Moon at Houston'. Reports by several hands. *Nature*, vol. 225 (24 January 1970), pp. 321-7.

MINING, MINERALS AND TECHNOLOGY

Books

Degens, Egon T. and Ross, David A. (Eds.), *Hot Brines and Recent Heavy Metal Deposits in the Red Sea*. Springer-Verlag, 1969.

Mero, John L., *The Mineral Resources of the Sea*. Elsevir Publishing Co., New York, 1965.

Papers and Articles

Deep Sea Drilling Project. Summaries, Drilling Reports etc. Issued by Scripps Institution of Oceanography, La Jolla, University of California, San Diego.

Kullenberg, B. 'The Piston Core Sampler'. *Svenska hydrogr.-biol. Kommn. Skr.* (ser. 3, Hydrog.), vol. 1, no. 2 (1947), pp. 1-46.

Pinam, Anthony, 'Deep Sea Drilling'. *Science Journal* (July 1970), pp. 61-67.

Proceedings of the British National Conference on the Technology of the Sea and Sea Bed, 5-7 April 1967.

Proceedings of the Ninth Commonwealth Mining and Metallurgical Congress, 1969.

Proceedings of Oceanology International, 1969 (Brighton).

PHYSICAL OCEANOGRAPHY AND METEOROLOGY

Books

Maury, Matthew Fontaine, *The Physical Geography of the Sea and its Meteorology*. (First published 1855.) Harvard Press edition 1963, Ed. John Leighly.

Stommel, Henry, *The Gulf Stream*. 2nd edition, University of California Press, Cambridge University Press, 1968.

Von Arx, W. S. *Introduction to Physical Oceanography*. Addison-Wesley Publishing Company Inc., 1962.

Papers and Articles

Baker, D. James Jr., 'Models of Oceanic Circulation'. *Scientific American*, vol. 222, no. 1 (January 1970).

Cartwright, D. E., 'Extraordinary Tidal Currents near St. Kilda'. *Nature*, vol. 223, no. 5209 (30 August 1969), pp. 928-32.

Cartwright, D. E., 'Tides and Waves in the Vicinity of St. Helena'. *Phil. Trans. Royal Soc. London* (in press, 1971).

Fofonoff, Nicholas P., 'Measurement of Ocean Currents'. Reprint 1969.

GARP Publications (Series) *No. 1 An Introduction to GARP* published by the World Meteorological Organisation and the International Council of Scientific Unions, October 1969.

Isaacs, John D., 'North Pacific Study'. *Journal of Hydronautics*, vol. 3, no. 2 (April 1969), pp. 65-72.

Isaacs, John D., Devereaux, R. F., Evans, M., Kosic, R. F. and Schwartzlose, R., 'Real Time Oceanographic Data from the North Pacific Study'. Paper read at International Telemetering Conference in Los Angeles, 9 October 1968.

Kuettner, Joachim, P., 'The Bomex Project'. *American Meteorological Society Bulletin*, vol. 50, no. 6 (June 1969).

Leighly, John, Introduction to 1963 edition *The Physical Geography of the Sea* by Matthew Fontaine Maury. Harvard Press.

Leighly, John, 'M. F. Maury in his time'. Paper read at 1er Congrès D'Histoire de Oceanographie, Monaco, 1966.

Munk, W. H., 'Once Again – Tidal Friction'. *Journal of the Royal Astronomical Society*, vol. 9 (1968), pp. 352-75.

Munk, W. H., 'Deep Sea Tides'. *Progress in Oceanography*, vol. 5 (1969), Pergamon Press.

Munk, W. H. and Zetler, B. D., 'Deep-sea Tides – a program'. *Science*, vol. 158 (17 September 1967), pp. 884-6.

Namias, Jerome, 'Seasonal Interaction between the North Pacific Ocean and the Atmosphere during the 1960's'. *Monthly Weather Review*, vol. 79, no. 3 (March 1969), pp. 173-92.

Snodgrass, F. E., 'Deep Sea Instrument Capsule'. *Science*, vol. 162, (4 October 1968), pp. 78-87.

Snodgrass, F. E., Groves, G. W., Hasselmann, K. F., Miller, G. R., Munk, W. H. and Powers, W. H. 'The Propagation of Ocean Swell across the Pacific'. *Philosophical Transactions of the Royal Society of London. Series A. Mathematical and Physical Sciences*, vol. 259, no. 1103 (5 May 1966), pp. 431-97.

Stommel, Henry, 'Future Prospects for Physical Oceanography'. *Science*, vol. 168, no. 3939 (26 June 1970).

Stommel, Henry, 'The Large-scale Oceanic Circulation'. Contribution to *Advances in Earth Science*, ed. P. M. Hurley, M.I.T., 1966.

Stommel, Henry and Arons, A. B., 'On the Abyssal Circulation of the World Oceans. I Stationary Planetary Flow Patterns on a Sphere' and 'II An Idealised Model of the Circulation Pattern and Amplitude in Oceanic Basins'. *Deep-Sea Research*, vol. 6 (1960), pp. 140-54 and 217-33.

Swallow, J. C., 'The Currents'. Chapter in *Oceans*, ed. G. E. R. Deacon (Paul Hamlyn, 1962).

Webster, Ferris, 'On the Representativeness of Deep-Sea Current Measurements'. *Progress on Oceanography*, vol. 5 (1969), Pergamon Press.

Webster, Ferris, 'Observations of Inertial Period Motions in the Deep Sea'. *Reviews of Geophysics* vol. 6, no. 4 (November 1968).

Webster, Ferris, 'Vertical Profile of Horizontal Ocean Currents'. *Deep-Sea Research*, vol. 16 (1969), pp. 85-98, Pergamon Press.

PERIODICALS AND OTHER SOURCES

Journal of Geophysical Research, American Geophysical Union, Suite 435, 2100 Pennsylvania Ave. N. W., Washington D.C. 20037. Three times a month.

Mondo Sommerso, Via Vigoni No. 11, Milano.

Nature, MacMillan Publishing Co., 4 Little Essex Street, London W.C.2. Weekly.

New Scientist, 128 Long Acre, London W.C.2. Weekly.

Oceans Magazine, Oceans Publishers Inc., 7075A Mission Gorge Road, San Diego, California 92120.

Petroleum Times, I.P.C. Business Press, 33-39 Bowling Green Lane, London E.C.1. Fortnightly.

Science, A.A.A.S., 1515 Massachusetts Avenue N.W., Washington D.C. 20005.

Science Journal, Dorset House, Stamford Street, London S.E.1. Monthly (specific back numbers).

Scientific American, Scientific American Inc., 415 Madison Avenue, New York, New York 10017. Monthly.

House Journals of: General Dynamics Corp.
                      General Electric Co.
                      Grumman Corp.
                      Lockheed Aircraft Corp.

Tenneco Oil Company

Westinghouse Electric Corp.

Marine Science Affairs, U.S., January 1969.

Scientific American – 'The Ocean' (special issue), vol. 221, no. 3 (September 1968).

The Stratton Report, 'Our Nation and the Sea'. U.S., January 1969.

Stone, P. B., 'A Policy for Oceanography'. *Penguin Science Survey*, 1968.

Acknowledgments

In a book of this kind, built around the work and opinions of so many people, it is hard to know where to begin in acknowledging help received. Having begun, it is equally difficult to know where to stop. For two authors the problems are not halved but doubled.

Let us begin then in the middle by declaring our joint debt of gratitude to those people in the United Kingdom who most particularly helped us in our task with advice and encouragement and also undertook the correction of finished chapters: Professor Sir Edward Bullard, F.R.S., of the Department of Geodesy and Geophysics, Cambridge; Rear-Admiral G. S. Ritchie, recently retired Hydrographer of the Navy, and his predecessor Rear-Admiral Sir Edmund Irving.

We are also grateful, together and separately, to many other people who have generously given their time to elucidating some of the mysteries of our subject-matter and tracking down essential facts, figures and references. Among them are Dr. Drummond Matthews and Dr. Dan McKenzie, also of the Department of Geodesy and Geophysics, Cambridge, and Dr. John F. Dewey, lately of the Department of Geology; Dr. John Swallow and Mrs. Mary Swallow, also Dr. David Cartwright and Dr. N. C. Fleming of the National Institute of Oceanography; Dr. P. L. Willmore of the Global Seismology Unit, Edinburgh; Miss Mary Deacon of the History of Science Unit, University of Edinburgh; Professor B. M. Funnell of the Environmental Studies Department of the University of East Anglia; Dr. G. P. L. Walker of the Department of Geology, Imperial College, London; Dr. Alan Charig of the British Museum, Natural History; Dr. R. W. G. Hazlett of Kelvin Hughes; Mr. G. A. Corby of the Meteorological Office, Bracknell; Dr. K. Bignall of the Department of Meteorology at Imperial College; Dr. F. R. Harden Jones of the Fisheries Research Laboratories, Lowestoft; Dr. Ian Richardson of the White Fish Authority and Mr. Keith Howard who was also of great assistance in this unusual field; Mr. Peter Hinde of the Gas Council and also Dr. J. S. Tooms for keeping speculation about brine sinks within proper bounds.

In the United States we are especially grateful for the help of the Oceanographer of the United States Navy, Admiral A. D. Waters Jnr., and for his permission to visit, in their laboratories, with the guidance of Dr. E. D. Schneider, some of the many scientists engaged in the U.S. Navy's Oceanographic research. We are also indebted to the United States Navy for research assistance from the Ships Histories Section of the Division of Naval Research.

At the Scripps Institution of Oceanography at La Jolla, California, we are grateful for the help of Professor H. W. Menard, Miss Tanya Atwater, Dr. J. D. Isaacs, Dr. T. Edgar, Dr. J. Morgan, Mr. Frank E. Snodgrass and many others; at the Lamont-Doherty Geological Observatory of the University of Columbia we valued conversations with Dr. W. C. Pitman, Dr. N. D. Opdyke, Dr. W. S. Broeker, Dr. Lynn Sykes, Dr. Billy Glass and John Foster. At the Woods Hole Oceanographic Institute we were assisted by Dr. Ferris Webster on the subject of current measurement and, on the tricky subject of foraminifera, by Dr. W. A. Berggren. Still in North America but moving to Canada, we are grateful for discussions with Dr. John Tuzo Wilson and his colleagues at Erindale College, Toronto, not forgetting the cheering sight of his pet budgerigar with a taste for gin and tonic.

Returning to the United States but in the commercial companies instead of the oceanographic institutions, we are grateful for the help of Mr. Bob McGrattan of Electric Boat; Mr. C. T. Coony of Vocaline Air Sea Technology; L. Cole and H. Arndt of Applied Oceanics, Maine; Mr. Willard Bascom of Ocean Science and Engineering; Arnold Rothstein and John Flipse of Deepsea Ventures, and to John L. Mero. We also thank

Messrs. Westinghouse, Lockheed, General Electric and General Dynamics for opening their doors and at least some of their files to us.

Before leaving the United States we also wish to remember the help of Mrs. N. Kaghan in tracking down people and of Mrs. R. Still in tracking down such things as photographs, maps and documents; also the hospitality of Mr. and Mrs. A. Proudfit and Mr. William Clark.

In Japan we had valuable help from Dr. Maniwa, Dr. Fujinaga and Mr. Izumi Yamashita, who almost persuaded Peter Stone to buy a fish farm.

Long before we arrived at the point of requiring help from so many people we had already acquired some relevant obligations. Brenda Horsfield is grateful to the British Broadcasting Corporation for the opportunity of producing, in 1966, the series of television programmes entitled 'The Unconquered Ocean', which led to meetings with so many oceanographers and incidentally laid the foundations of this co-operation.

Peter Stone also wishes to express his thanks to those who taught him geology, especially the late Professor L. R. Wager, Dr. W. S. Mackerrow, and Dr. K. S. Sandford at the University of Oxford, the last two of whom have been kind enough to retread some of that ground with him in recent conversations, as did Dr. Harold Reading.

Finally together again, we join in thanking Miss Elizabeth McDowell, Miss Caroline Stott and Miss Jane Hoyle, who typed our manuscript with such care.

ILLUSTRATIONS

We are grateful to the editors of *Scientific American* for permission to copy the drawing on page 307; to the editors of *Nature* for permission to reproduce the diagrams on pages 101 and 167, and to the editors of *Science Journal* for permission to reproduce the illustrations on pages 105, 125, 138, 142, 143. With the exception of certain maps, all other illustrations have been specially drawn for this book, the majority by DAVID PENNEY. Diagrams based on figures published in scientific papers or books are, where the original source is not made clear in the accompanying caption, acknowledged as follows:

page
36 Polar curves – Runcorn, 1962
56-57 Guyots – Hess, 1946
60 Afar triangle – Tazieff, 1970

Index

boundary layer, 311
brine sinks, 169, 170, 194, 249, 257 ff.
British Columbia, 93, 131
British Isles, 149, 154, 293
British policy, lack of, 186, 189, 192-8
Britanny, 293
Bronze Age, 152
Brooke, Midshipman J. M., 41, 42
Brown and Root, 211
Brunel, I. K., 42
Brussels, 294, 296
Buchanan, 47
Bullard, Prof. Sir Edward, F.R.S., 69, 79,
    110, 126, 147
bumper sub, 210
buoys, 69, 291, 313 ff.
    monster buoys, 309 ff.
Burke, William T., 320
Burma, 168
Burstyn, Harold L., 45

cables (telegraph), 43 ff., 50
Caledonian period, 26
Caledonides, 131, 133
California, 64, 98, 117, 131, 151, 285
    Gulf of, 116
Cambridge, 22, 95, 98, 100, 133, 168
Cammell Laird, 228-30
Canada, 97, 98, 135
Canadian shelf, 141
Cape Johnson, U.S.S., 56, 63
Cape Johnson Deep, 63
Cape Town, 151
Cape Verde Is., 171
capsule (tide), 290 ff.
carbon, 166
Carboniferous period, 32, 34, 88
Carey, S. Warren, 81
Caribbean, 105, 125
Carlsberg Ridge, 55, 74, 95, 96
    spreading rate, 103, 124, 140
Carnegie, 55
Carnegie Institute, 106
Carpenter, Sir William B., 45
Cartwright, David E., 289, 293
Cascade Mountains, 79, 148
casing (drill), 205
catalysts, 163, 164

'catastrophe', 82, 83
cells, 163, 165; *see also* convection
CEMA, 188
Cenozoic era, 121
Ceylon, 166 ff.
Challenger, H.M.S., 12, 42, 43 ff., 55, 59,
    61, 63, 72, 249
Challenger Expedition (1872-75), 12, 43-50,
    296
Challenger Report, 48, 49
Chandler wobble, 109
Charig, Alan, 157
chemical evolution, 163, 165
chemicals (toxic), 305, 308
chemistry, 162 ff.
chert, 67, 105
chess, 125
Chicago, University of, 163
Chile (Southern), 74
chlorella, 275
Christmas tree, 207, 220, 225
circulation, 297 ff., Ch. 17
circumference of earth, 80, 81
Circumnavigation Committee, 46
clamm, 40, 41
clams, 276
Clarion fracture zone, 64
climatic change, 157, 160
Clipperton fracture zone, 64, 65
Clipperton Island, 209
C.N.E.X.O., 187-8, 196
Cockcroft, 12
cockles, 276
collision, 138, 140, 144, 146, 155
Colorado River, 71
colour differences, 160
Columbia University, 68
COMEX, 187-9, 217
compressed air gun, 68
computers, 71, 76, 96, 102, 286
    fit, 110, 168, 288, 289, 296, 311
Concorde, 197
condenser discharge, 55
congruent shape, 87
conservation, 323
Conshelf, 218
Constantinople, 159
continental drift, 28 ff., 78, 80
    dates of, 112, 149, 150, 154 ff., 161

plastic layer, 123, 144
plate consumption, 135 ff.
plate tectonics, 113, 134, 135, 136, 159, 166
plates, 19, 113 ff., 135 ff., 162
  moon, 165
  planets, 166
Plato, 77, 78, 152
polar (North) ice cap, 74
polar wandering curves, 149
pole of rotation, 118 ff.
pole of spreading, 118 ff.
pollution, 191, 290, 314-18
polyp, 58
pools, 169
Porcupine, H.M.S., 45
potassium-argon, 95, 100
prawns, 281-3
precision depth recording, 62
pressure gradients, 299
primitive earth, atmosphere, 163
Princeton, 56, 57, 81, 97, 99
propane, 309
protein, 163, 306
proto-Atlantic, 131
provinces, faunal, 157
pseudo-accretion, 137 ff.

Q-ships, 53
quartz, 163, 164
Quebec, 98
Queen Charlotte Islands, 93
  fault, 101, 116

race, 160
Racehorse, H.M.S., 40
radioactive dating, 36
radioactivity, 82
radiocarbon dating, 301
radiolaria, 67, 106, 160
Raff, Arthur, 94
Ramapo, U.S.S., 61
Rand, John R., 131, 132
Rattlesnake, H.M.S.S., 44
Red Sea, 19, 59, 72, 76, 117, 149, 194, 249, 257-8
reflection, acoustic, 53
  seismic, 68, 69
refraction, seismic, 68, 69, 70

reptiles, 155 ff.
research ships, 50
research ship techniques, 40
research submarines, 232
Revelle, R. R., 79
reverse magnetisation, 94 ff.
  cores, 106
rhinoceros, woolly, 161
Rhodesian Craton, 141
Richards, Vice-Admiral Sir George, 43, 45, 46, 47
Richter, C. F., 71
Richter scale, 109
ridges, mid-ocean, 77
  crests, 113 ff., 134 ff., 144 ff., 169
rift valleys, 59, 71
rifting, 157
'ring of fire', 27, 73
riser pipe, 207, 210, 215
Rockall, 154
Rockies, 131
Rodius, 22
Roekall, 319
Ross, James Clark, 41
Ross, John, 40, 41
rotation on sphere, 118 ff.
Rothé, J. P., 72
round trips, 202
Royal Navy, 194, 218
Royal Society, The, 45, 47, 49, 58, 110, 293
Royal Society of Canada, 98
Runcorn, Prof. S. K., F.R.S., 88, 126, 130
Russia, 51
Russian oceanography, 189-90, 325
Rutherford, Sir Ernest, 53

sabretooth, 161
St. Helena, 76, 289
St. Peter and St. Paul Rocks, 76
salinity, 295, 301
salmon, 278, 323
salt domes, 171
Salton Sea, 258 ff.
Salvation Army, 49
Samoa, 92, 285
San Andreas fault, 100, 101, 116 ff.
San Diego, 92
San Francisco, 116, 117
Santorini, 152